Manabu Kanda

translated by

Alvin C. G. Varquez

Ordinary Differential Equations and Physical Phenomena

A Short Introduction with Python

常微分方程式と物理現象

神田 学 [著]／アルビン・バルケズ [訳]

Asakura Publishing

First published in Japan in 2020
by Asakura Publishing Company, Ltd.
6-29, Shin'ogawa-machi, Shinjuku-ku,
Tokyo 162-8707, Japan

ISBN 978-4-254-20169-7

Preface

Except for students majoring in mathematics, students enroll in various science and engineering fields because of their curiosity of understanding engineering and physical phenomena, more than their interest in mathematics. I frequently wonder and would like to believe that the career choices of such individuals are indeed based on their fascination with the various physical phenomena around us. With this mindset, I feel that the mathematics courses offered at the undergraduate level focus mainly on solving pure math problems with little discussion on how to relate the solutions to physical or natural phenomena. This style of education is comparable to how a language is taught. In linguistics, thoroughly mastering grammar and vocabulary precedes the mastery of techniques for constructing interesting, practical conversations. Nonetheless, the value of mathematics reveals itself when the students choose to specialize in a certain field. Only then will the student see the value of the mathematics course which they learned earlier in their undergraduate life. In this textbook, I want to balance the importance of physical phenomena and mathematics, with the latter as a tool for understanding the mechanisms of the former. Meanwhile, I want to encourage the reader by using simple and understandable explanations of concepts and solutions. Therefore, this textbook is quite different from existing mathematics books that deal with ordinary differential equations. Furthermore, Most undergraduate-level mathematics textbooks in Japan focus on problems that are solvable by hand. Numerical methods that are very useful for solving non-linear differential equations are not tackled in most textbooks. Based on the assumption that "the language presented by ordinary differential equations is universal since it can represent a wider range of physical phenomena" presented in this textbook, exact and numerical solutions of linear and non-linear phenomena ought to be discussed. Therefore, the aim is to have a broad overview of "ordinary differential equations." As a trade-off, this textbook omits sophisticated solutions rarely used in specialized fields of science and engineering, clever solutions that limit the imagination of physical phenomena, and even mathematical proofs.

As a specialist in urban environmental science, it has been a daunting task for me to write this mathematics textbook. What motivated me to pursue writing this textbook is the fact that our society is entering a more globalized environment, with knowledge advancing rapidly both domestically and internationally. We have to increase our momentum of educating our youth who can bridge or see a common ground among highly specialized fields. Also, I am a faculty member of the Department of Transdisciplinary Science and Engineering of the Tokyo Institute of Technology. Uncommon for Japanese universities, this department advocates the globalization of our students and exposing them to various fields of science and engineering. To contribute to this unique program, a textbook is needed for second-year students to understand mathematics as a common language tool for understanding various physical phenomena. This textbook contains exercises including programs for visualizing and understanding physical phenomena based on the underlying ordinary differential equations which describe them. I hope you can use them in your lectures for undergraduate students, or even for self-study.

I would like to express my gratitude to Asakura Publishing. I also would like to thank my students and assistant professor, Dr. Atsushi Inagaki, for checking the solutions of the exercises. Finally, I would like to thank my family for giving me inspiration and support.

February 2012

Manabu Kanda

Professor
Department of Transdisciplinary Science and Engineering
Tokyo Institute of Technology

Preface to the English Edition

Back in 2016, I began teaching the course "Ordinary Differential Equations and Physical Phenomena" for undergraduate students of Tokyo Insitute of Technology. For preparation, I sat through the same lecture conducted in Japanese by Prof. Manabu Kanda. It was during that time when I encountered his enlightening book entitled "常微分方程式と物理現象, 神田学, 朝倉書店" which he used as a guide for the lecture. I was deeply impressed by the book's structure and style that I felt the need to share its content with everyone. After my first semester of teaching, I completed rough translations of various sections of the book for my students. After five years of teaching, a fully translated version was completed with additional program-based exercises in Python and other discussions.

This book is not a typical math book that focuses on problem-solving techniques. The main objective of this book is to provide a clear purpose of why mathematics is useful to explain various physical phenomena surrounding us. The book serves to aid students in engineering and other science-related majors by bridging elementary or foundational mathematics to their field of interest. It also encourages the use of mathematics as a communication tool across disciplines.

Exercises are introduced to visualize physical phenomena represented by the differential equations. In this translated version, Python is the programming language used for the exercises. The programming language is a widely used tool by engineers, scientists, and graduate students in rapid calculations, software-building, and analyses. It is also freely downloadable and compatible with most platforms. I thought it as an added learning bonus for readers to expose themselves to programming.

It is my wish that readers will appreciate mathematics; and will not be troubled by the question, "Why do I need to learn differential equations and how is it even relevant to my career or life?" After appreciating mathematics through this book, readers may pursue higher levels and deeper mathematical subjects.

I especially thank the editors of Asakura Publishing for their time, effort, and patience during the preparation of the book. I am grateful to my first teaching assistant, Dr. Muhammad Rezza Ferdiansyah. Without his help, I probably would not have had the time to translate this book. I deeply thank Prof. Manabu Kanda and his laboratory for their constant support and encouragement. As my former doctoral supervisor and mentor, Prof. Kanda continues to inspire me to be a better researcher and teacher. I am especially grateful to the Global Scientists and Engineers Program (GSEP) of the Department of Transdisciplinary Science and Engineering for granting me the opportunity to teach and for bridging Japanese education to the world. I also thank my former talented students for their feedback and active participation. Finally, I dedicate this work to my family, friends, and God.

September 2020

Alvin C. G. Varquez

Associate professor
Department of Transdisciplinary Science and Engineering
Tokyo Institute of Technology

Table of Contents

1 Overview   **1**

  1.1 Ordinary differential equations as an engineering language   1

      1.1.1 Engineering and ordinary differential equations . .   1

      1.1.2 Definition of a differential equation 4

      1.1.3 Classification and terminology 7

      1.1.4 The first solution: The basic concept 12

  1.2 Very basic Python tutorial 20

2 1st order ODE   **23**

  2.1 Variable separation method 23

  2.2 Variable separation of 1st derivatives of linear expresssion   23

      2.2.1 Exact solution and physical meaning 23

      2.2.2 Free-falling body 25

      2.2.3 Self-propagating/self-destruction of microorganisms in a container 26

      2.2.4 Electric circuit composed of resistor and capacitor: RC circuit 27

      2.2.5 Decay of radioactive elements 28

      2.2.6 Attenuation of solar radiation (Beer's law) 29

      2.2.7 Balance of physical quantity in space: Box model . 30

  2.3 1st order non-linear ODE solvable by variable separation method 32

      2.3.1 Population dynamics: Logistic curve 32

  2.4 Exercise 1 33

      2.4.1 Investigating propagation of species 33

      2.4.2 Procedures for solving Exercise 1 34

3 2nd order linear ODE   **39**

  3.1 Method to deriving an exact solution 39

  3.2 Mass–spring system 43

  3.3 Behavior of current flowing in an RCL circuit 48

  3.4 Disk rotation subjected to tortion and resistance 49

  3.5 Deriving the law of conservation of energy using the law of conservation of momentum 50

      3.5.1 Derive the law of energy conservation from law of momentum conservation 50

        3.5.2   Derive the equation of motion from the law of en-
                ergy conservation   51
    3.6 Exercise 2   52
        3.6.1   Investigating the behavior of a mass–spring system
                without external influence   52
        3.6.2   Procedures for solving Exercise 2   53

4   2nd order linear non-homogeneous ODE   60

    4.1 Deriving the exact solution   60
    4.2 Behavior of mass–spring system with periodic external force   62
        4.2.1   Deriving a general solution of linear homogeneous
                ODE   63
        4.2.2   Particular solution of linear non-homogeneous ODE   64
        4.2.3   Deriving an exact solution of linear non-homogeneous
                ODE   65
    4.3 Behavior of current in RLC electric circuit with AC voltage   68
        4.3.1   Deriving a general solution of linear homogeneous
                ODE   69
        4.3.2   Particular solution of linear non-homogeneous ODE   70
        4.3.3   Deriving a general solution of linear non-homogeneous
                ODE and its practical resonance   70
    4.4 Exercise 3   73
        4.4.1   Investigating the behavior of a mass–spring system
                with external influence   73
        4.4.2   Procedures for solving Exercise 3   74

5   Numerics for ODEs   85

    5.1 Basic concept of the solution   85
    5.2 Euler method   86
    5.3 4th order Runge–Kutta   89
    5.4 Numerical solution of multiple simultaneous 1st order ODEs   92
    5.5 Numerical solution of n-order ODEs   93
        5.5.1   2nd order ODEs   93
        5.5.2   3rd order ODE   94
    5.6 Exercise 4   95
        5.6.1   Applying the Euler and Runge–Kutta to the logistic
                model   95
        5.6.2   Applying the Euler and Runge–Kutta to the mass–
                spring system   95
        5.6.3   Procedures for solving Exercise 4   96

6   ODE and chaos   103

    6.1 Non-linear mass–spring system and chaos   103
    6.2 Physical definition and characteristics of chaos   104
    6.3 Characteristics of chaos in the non-linear mass–spring system 106

6.3.1   Multi-periodic cycles   107

6.3.2   Extreme sensitivity to the initial conditions   107

6.3.3   Chaos comes from non-linear internal systems   . .   109

6.3.4   Strange attractor in the non-linear mass–spring system   110

6.4   Food chain and chaos   111

6.4.1   Population of predator and pray: Lotka–Volterra equation   111

6.4.2   Subordinates (prey), mid-rank organisms (predator-prey), top organism (predator): Chaos phenomenon from co-existence of 3 organisms types   112

6.4.3   Chaos in food web models   114

6.5   Exercise 5   119

6.5.1   Lorenz model   119

6.5.2   Procedures for solving Exercise 5   120

Answers to Problems   **124**

A   List of formulae   **141**

A.1   Trigonometric identites   141

A.2   Differentiation and integration formulae   142

Index   **143**

1 Overview

1.1 Ordinary differential equations as an engineering language◦1

◦1 This section aims to answer questions such as "How is mathematics related to the building of robots?"

1.1.1 Engineering and ordinary differential equations

◆ Cross-sectional view of engineering

Engineering aims to acquire and provide scientific understanding regarding natural or physical phenomena, from which tools necessary for mankind's survival are developed. Throughout decades, engineering has branched out to various fields. Mechanical engineering, electrical engineering, civil engineering, chemical engineering, and other engineering disciplines were established to provide in-depth focus on specific physical phenomena. Despite the seemingly different nature and application of these engineering fields, their mathematical foundations and forms are analogous with each other.

Until the beginning of the Meiji era, education in Japan, as with its neighboring Asian countries, was more broad and generalized into three main areas: humanities, craftsmanship, and nature. After the arrival of Western influence, the concept of "specialization" and "division of labor" became an efficient strategy for the nation's advancement. Henceforth, individual roles became more distinct and the main fields◦2 were further subdivided into specific subfields. Although it proves efficient, this fragmented perception limits an engineer to understand complex systems such as crises on a global scale. To put it simply, the side-effect of specialization of field is a degradation of communal knowledge among citizens.

◦2 In the field of medicine, can you enumerate various specializations?

In the modern times when global issues have become more complex and wider in scale, the ability to communicate across areas have become indispensable. This book is dedicated to introducing ordinary differential equations (ODE) as a mathematical tool (differential equations) which aids in understanding physical phenomena deeply and to re-establish links among various fields or disciplines of engineering.

Differential equations can be any equation that is comprised of the derivatives of mathematical functions commonly used in modeling physical phenomena. In this book, ODE is the primary focus. ODE and its application will be taught in a very general manner without focusing too much on the various techniques for solving them (since these techniques can be readily acquired from any textbook on advanced mathematics). Upon defining ODEs and recognizing its purpose, the reader will be able to apply this to various fields of engineering and revive the communal knowledge that is at the risk of being obsolete.

◆ ODE as a "universal" expression for physical phenomena[3]

An analogy to explain the relationship between physical phenomena and ODE is shown in Figure 1.1. The type of trees according to ecology can be recognized through the foliage (plant cover) and historical records of a certain tree's appearance. In addition, nomenclature of trees also vary by language. For example, the word "tree" is called "jumoku" in Japanese or "Baum" in German. The countless tree types multiplied by its various translations would require a specialist to spend a lot of time memorizing, cataloging, and recalling common features of certain trees. On the other hand, describing a tree mathematically can be much simpler by summing up behavior or patterns of various physical phenomena (by mathematical expressions) occurring both internally and externally from the tree. This concept can be applied to any system or body; and even to the smallest of species. With careful inspection through experimentation or modeling, the descriptive parameters can be acquired and may serve as a tool for classifying almost anything.

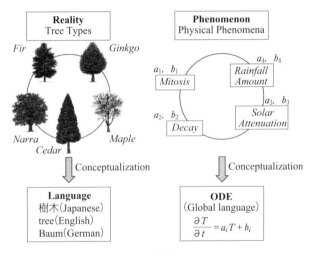

Figure 1.1   Verbal Language vs. ODE
Left figure taken from http://www.wood.co.jp/mz/index.html

Let us take the case of a butterfly. The purpose of natural history is to classify and organize the diversity of butterflies, which includes details down to the microscopic level[4]. Natural history captures diversity from universality. On the contrary, essential and common features are described by differential equations in the field of engineering[5]. The current situation is that each engineering disciplines are set apart despite the common essence in the problems being faced.

It can be stated that differential equations can serve as a universal (or

⇨**3**   It seems other departments teach similar subjects such as industrial engineering mathematics. Is there any difference? Only in appearance. Mathematics taught in engineering only varies in the focused phenomenon. Essentially, they are the same.

⇨**4**   In detailed descriptive analysis such as those used in encyclopedias, mathematical expressions are not as useful as in engineering applications.

⇨**5**   In engineering and physics, finding universality from various realities is a major task, so mathematics is highly recognized as a useful tool.

global) language which describes various physical phenomena. Through ODEs, various trees can be classified according to quantifiable parameters. In the field of engineering, various physical phenomena can be described simply from differential equations describing Newton's laws. Mastering mathematical expressions such as differential equations proves to be useful to eliminate the need for memorization to a certain degree.

In modern-day usage of engineering, diversity becomes a starting point in order to define a more universal concept or behavior. Complex issues faced in society can be understood when experts not only share their inputs but also interpret and respond to the input of others. Dissecting the parameters and understanding the fundamental concepts of each fields of expertise, an engineer equipped with sufficient mathematical skills understands the similarity across engineering fields.

Mathematical description of phenomena is not the end objective of engineering. If the differential equation is translated analytically or approximated sufficiently, the physical phenomenon conceptualized by mathematical formulae can be virtually reproduced. This translation of mathematical formulae is referred to as "solving differential equations." The role of engineering is not only to express or define physical phenomena mathematically or universally. With the aid of computing technology, case studies by experimentation or modeling can be conducted in order to understand a system's behavior under certain parameters or conditions⟜**6**. These investigations, under acceptable uncertainties, can be used to remedy various issues or problems in the future.

⟜**6** By looking at weather forecasts, it can be deduced that weather as a complex system is difficult to reproduce in laboratories. In recent years, numerical methods (ch. 5) to derive approximate solution of ordinary differential equations using a computer are used to determine relevant parameters and initial values. Experiments using a computer is called a numerical experiment.

In linguistics, a good memory is required to remember multiple terminologies. Memorization is still necessary in engineering only to a certain point of being able to recall mathematical concepts. The terminologies used in ODE do not require much memorization. On the other hand, the most strenuous mental activity would be in understanding the mathematical concepts. Nevertheless, with time and continual practice, mastering the concepts of ODE can provide the wielder with a convenient yet powerful tool.

◆ **Mutual relationship between formulas and physical phenomena**⟜**7**

⟜**7** Students tend to ask questions such as "Is it necessary for me to be able to solve typical problems to pass the university entrance exams?" and "Which formulae should I memorize?" Occasionally, examiners provide the mathematical formulae. This means, how to use the given mathematics efficiently as a tool matters more than the formulae themselves.

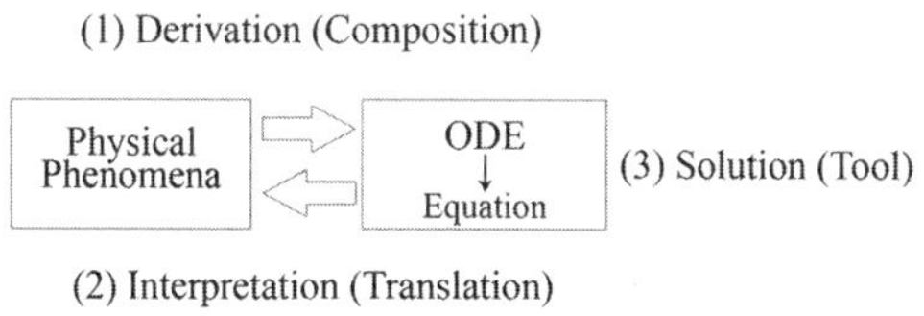

(1) Derivation (Composition)

Physical Phenomena ⇨⇦ ODE ↓ Equation (3) Solution (Tool)

(2) Interpretation (Translation)

Figure 1.2 Key to studying ODE in engineering

The study of ODE is a repetition of 3 steps (Figure 1.2):

1. **DERIVE** the ordinary differential equations from the phenomena.
   (Composition: Physical phenomena → Equation)

2. **INTERPRET** the physical meaning displayed by ODE.
   (Translation: Equation → Physical phenomena)

⇨8 Numerical analysis (ch. 5) can be said to be a universal solution. With enhanced computational libraries, engineers can solve problems faster.

3. Understand the basic **SOLUTION** (learning tool⇨8).

In conventional methods of mathematics education, it is more common to focus on the techniques/equations to immediately reach a solution than the derivation of the differential equations that describe the physical phenomena or problem of interest. Following conventional approach makes interpretation of the solution difficult and limits the flexibility to solve multiple problems. Applying the solution derived by conventional methods to more complex problems is not possible and the student will not acknowledge the significance of ODE and will forget it altogether.

It is through the steps, (**derive**) and (**interpret**), that one can universally explain physical phenomena and link it with mathematics via basic **solutions**. By repeating the self-explanatory steps above, ODE will become an automated and iterative tool.

As will be discussed later, there are different ways to solving an ODE that either provide an exact or approximate solution. The exact solution of ODEs are valuable only as an education tool capable of solving idealized systems. Problems requiring exact solutions are far from applicable in the real world. With computational advancement, approximate solutions which can be applicable for various complex systems can be acquired and set to closely match exact solutions.

In the modern times, the level of mathematical skill required for an engineer is that which can allow deeper understanding of unknown or unexplained physical phenomena. Throughout this book, the steps to studying ODE, mentioned above, will be followed. Specifically, deriving appropriate equation to represent the phenomenon, setting up boundary and initial conditions, and interpreting the results physically, will be conducted.

Memorizing theorems (or postulates) is not necessary in this textbook and will be left to the student for further review or practice. The reason behind this is that this textbook is not a means to improve memorization skills but critical thinking skills and appreciation. There is zero merit for your brain to memorize theorems without knowing its usage. As an analogy, owning an extensive and powerful computer is useless if the owner only uses it for office applications!

1.1.2   Definition of a differential equation

◆ Dependent and independent variables

Table 1.1 Variable types and notations

Variable	Notation	Other Name	Physical Quantity	Physical meaning
Dependent Variable	T	Explained Variable	Concentration, temperature, displacement, etc.	What we want to know
Independent Variable	x, y, z, t	Explanatory Variable	Space, time	ID/Address/Index

Differential equations are comprised of variables and mathematical expressions. It is of primary importance to identify the variables and their types within and externally influential to a given physical system.

The types of variables and their physical meaning is demonstrated using an example. Imagine determining a room's temperature for a certain time. The necessary variables are temperature and time. Let us denote temperature and time as T and t, respectively. Notations (e.g. letters, characters) that expresses variables are arbitrarily selected for convenience. By logic, the variables, T and t, are changing within the system (i.e. room). Furthermore, both variables are related to each other such that we can set T to be a function of t (i.e. temperature changes with time). To put it simply, temperature depends on the value of time, whereas time is assumed to vary independently with other variables. T and t can therefore be classified as a dependent and an independent variable, respectively (1.1). Dependent variables can be written as functions of independent variables. For this case, T can be written mathematically as $T(t)$. Dependent variables (or explained variable) could be anything which describes the state of the system of interest (e.g. concentration, population, displacement, velocity). The independent variable (i.e. explanatory variable) may be any physical quantity that can be expressed mathematically (e.g. space and time). Both dependent and independent variables are essential components of an ODE.

With regards to the earlier example, another approach to determine the room temperature aside from ODE would be to construct empirical functions such as those from statistical regressions. For example, a solution can be derived based on a derived empirical relationship (based on historical information) between the average temperature of the room and a city's temperature or a city's power consumption (e.g. measurements from agencies). A typical approach is linear regression analysis[9]. This approach, although still mathematical, tends to simplify the problem and would require long-term availability of measurements of a city's temperature, power consumption data, and the room's information. The major limitation of empirical methods is that the solution is not necessarily applicable universally[10]. Likewise, the city's temperature and power consumption are also dependent on other variables that are unique per city.

Rather than statistical or empirical approach, engineers aim to derive a solution that can express the phenomena for any given time and space (i.e. universal solution). Instead of adding power consumption or city temperature to the independent variables, spatial variables (x, y, z) could be used in addition to t. The most common spatial coordinate system is the Cartesian coordinate system where points are linearly spaced at certain intervals. Other coordinate systems can be used as well such as polar or cylindrical coordinate systems. When Fourier transform[11] is applied to the variables, the spatial variables, x, y, and z, are replaced with angular frequency as independent variables.

Independent variables are also convenient in describing the state of the dependent variable in a specified time or space. If, for example, spatial average temperature is not required but rather a specific temperature for a given time and space, T may be further expressed by $T(x, y, z, t)$ or indexed variable such as $T^t_{i,j,k}$.

◆ Definition of "derivative"

In addition to variables, differential equations also contain derivatives.

▷**9** Linear regression analysis is approximating the dependent variable T by the linear sum of multiple independent variables $t_1, t_2, t_3, ...$
$$T = a_1 t_1 + a_2 t_2 + \cdots + a_n t_n$$

▷**10** In this textbook, we separate the usage of differential equations for physical modeling and statistical modeling. The focus is on modeling physical phenomena. As a physical model, dimensions (or units) of all terms of the equations must be the same.

▷**11** Not covered in this textbook. The dependent variable can be represented by the sum of trigonometric functions of varying frequencies called Fourier series. This is closely related to a mathematical operation called Fourier transform which decomposes signals to frequencies, or converts the time-domain function to a frequency-domain function. Its reverse process is called the inverse Fourier transform.

Table 1.2 Physical meaning of the derivative

Order	Equation	Abbreviation	Definition	Physical Application
1st Order	$\frac{\mathrm{d}T(t)}{\mathrm{d}t}$	$T'(t)$ T_t	$\lim_{\Delta t \to 0} \frac{T(t+\Delta t)-T(t)}{\Delta t}$	Speed (time) Slope (space)
2nd Order	$\frac{\mathrm{d}^2 T(t)}{\mathrm{d}t^2}$	$T''(t)$ T_{tt}	$\lim_{\Delta t \to 0} \frac{T'(t+\Delta t)-T'(t)}{\Delta t}$	Acceleration (time) Roughness (space)

Derivatives are expressions corresponding to the rate of change of a dependent variable with respect to an independent variable. "To differentiate" in calculus and ODE means to perform a derivative. Mathematically, the derivative can be estimated by the limit of the ratio of the change of T from t to $t + \Delta t$, and Δt, as Δt approaches zero (see Table 1.2). If T is a variable or function of an independent variable, T' can denote the **1st derivative** of the variable or function. Differentiating T' leads to the **2nd derivative** of T denoted by T''. The **n-th derivative** of T with respect to an independent variable is also called the **derivative of order n**. In more advanced mathematics, you will encounter higher-order derivatives which basically mean repeated derivatives of functions or variables.

Given a function of a curve (e.g. $T(t)$), the slope of a tangent line can be mathematically represented by the 1st derivative. In theory, the 1st and 2nd derivative of displacement with respect to time represents speed and acceleration, respectively. On the other hand, the derivatives of displacement with respect to space could represent slope (1st derivative) and concavity/roughness (2nd derivative)[12].

Since the derivative represents a change of a certain variable, calculating the difference or incremental change of a dependent variable for a given interval of the independent variable using a computer can be used to approximate the derivative (to be discussed in the future chapters).

[12]

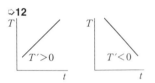

The first derivative represents the gradient (e.g. velocity) as shown above.

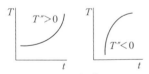

The second derivative represents the roughness (acceleration) of the dependent variable as shown above. Left shows acceleration (concave). Right shows deceleration (convex).

◆ Ordinary vs. partial differential equations

Equations that include a combination of dependent and independent variables are called differential equations. Differential equations with only one independent variable is called an **ordinary differential equation** (ODE), those with two or more independent variables are called **partial differential equations** (PDE). The number of independent variables indicates whether a differential equation is an ODE or PDE. Given that time, t, is assumed to be the sole independent variable influencing T for the first example, the differential equation which can be formulated is an ODE (eq. 1.1).

$$
\begin{aligned}
\textbf{ODE} \quad & f\left(T, \frac{\mathrm{d}T}{\mathrm{d}t}, \frac{\mathrm{d}^2 T}{\mathrm{d}t^2}, t\right) = 0 \\
\textbf{PDE} \quad & f\left(T, \frac{\partial T}{\partial t}, \frac{\partial^2 T}{\partial t^2}, \frac{\partial T}{\partial z}, \frac{\partial^2 T}{\partial z^2}, \frac{\partial^2 T}{\partial t \partial z}, t, z\right) = 0
\end{aligned}
\tag{1.1}
$$

Another obvious indicator to identifying whether a differential equation is an ODE or PDE would be to identify the presence of the symbols d (ordinary) and ∂ (partial).

What if the CO_2 concentration, C, in addition to T of the room is to be determined? In this case, the equation will have two dependent variables and an independent variable, t. Since independent variable remains singular, equations comprised of both C and T are still classified as ODE. However, it becomes necessary to state two functions (f_1, f_2) (eq. 1.2). These sequence of ODEs are referred to as simultaneous ODEs[13].

$$
\textbf{Simultaneous ODE} \begin{cases} f_1\left(T, C, \frac{dT}{dt}, \frac{d^2T}{dt^2}, \frac{dC}{dt}, \frac{d^2C}{dt^2}, t\right) = 0 \\ f_2\left(T, C, \frac{dT}{dt}, \frac{d^2T}{dt^2}, \frac{dC}{dt}, \frac{d^2C}{dt^2}, t\right) = 0 \end{cases} \tag{1.2}
$$

Simultaneous ODEs can be solved following basic procedures. On the contrary, simultaneous PDEs are much more complicated. Its complexity dramatically increases with increasing number of independent variables.

1.1.3 Classification and terminology

◆ Order of ODE: Required boundary conditions, initial conditions

An ODE is said to be of the order n if the highest rank among the derivatives included in the ODE is n. For example, $f = T'' + T' + T = 0$ is a 2nd order ODE. Integrating f will result in a new equation of one order lower and a constant. Further repeating the process up to the 0-th order, the n-th leading derivative disappears and the number of unknown constants increases to the number of times integration was conducted[14]. The final function comprised of the dependent variable, independent variable, and all unknown constants is called the **general solution** of the ODE. Assigning arbitrary values to the unknown constants will lead to multiple unique solutions. A unique solution is commonly called a **particular solution** [15].

In reality, the values assigned to unknown constants of the general solution should be consistent with the physical phenomenon of interest. In order to determine the n number of unknown constants, n number of expressions representing various physical conditions are required. A 2nd order ODE where the independent variable is time (t) and the dependent variable is displacement (X) and its 1st derivative is velocity (dX/dt), its 2nd derivative, acceleration (d^2X/dt^2), will require certain number of known conditions. To determine a realistic value for the constant of integration, an initial $(t = 0)$ value for displacement and speed is necessary. This is called an **initial condition**. In the case where the independent variable is in spatial coordinates, values of the dependent variable at the boundaries are required. This is called a **boundary condition**. Depending on the initial and boundary conditions, n unknown constants are physically determined and the functional form of the dependent variable will have an **exact solution** specific to the corresponding physical phenomenon. The rule is that in order to obtain an exact solution of any given ODE of nth order, n number of initial and boundary conditions are required. Engineers have to be very careful in determining the initial and boundary conditions if the target physical phenomenon is to be accurately represented.

◆ General solution, particular solution, exact solution

A solution of an ODE can either be general, particular, or exact. Let us suppose that the derivative of a dependent variable (1st order ODE) is given[16]. The relationship of the three types of solutions are shown in Figure 1.3. Integrating the differential equation, $T' - 3t^2 = 0$, leads to a

▷**13** As described earlier, the fundamental meaning of problem solving is to reduce difficult problems into simple ones. Even though solving PDEs are more complicated than ODEs, the general procedure is the same. In fact, many exact solutions to PDEs are simplified in terms of ODEs.

▷**14** This is for cases when integration is actually possible and exact solution can be obtained. For other cases, an approximate solution can be obtained using numerical analysis. Nevertheless, the relationship between the highest order of the derivatives and the necessary number of boundary and initial conditions remains the same in principle.

▷**15** A particular solution is obtained when a specific value is assigned to the constants of integration.

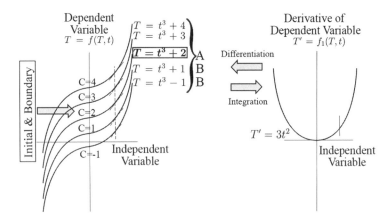

Figure 1.3   Relationship between functional forms (general solution, "A"; particular solution, "B"; and exact solution, boxed) and first derivative

⟳**16**   A simple example of a 1st order derivative is shown in Figure 1.3.

⟳**17**   See the general solution in Figure 1.3. Since the integration of the derivative does not explicitly define the constant of integration, an infinite number of solutions can exist. This is represented by a general solution.

⟳**18**   An exact solution is obtained when specific values described by the initial and boundary conditions are substituted to the constants of integration.

⟳**19**   Although actual values of the constants are different, the slope of the tangent line, as shown by the plot of the derivative on the right, is the same.

family of solutions[⟳17] in the form $T = t^3 + C$, where C corresponds to an infinite number of possible values. The form of the family of solutions is represented by a **general solution**, a singular expression characterized by an arbitrary constant of integration (C is used for the example). Based on known initial and boundary conditions taken from actual physical conditions, an **exact solution**[⟳18] can be obtained where C is replaced with a constant (i.e. since $T = t^3 + 2$ in Figure 1.3 is an exact solution, the constant "2" must be based upon the initial and boundary conditions). On the other hand, substituting any value of C (e.g. $C = -1, 1, 2, 3, 4$) corresponds to a **particular solution** (i.e. a solution without any arbitrary constants[⟳19] and possibly no physical meaning).

◆ **Linear and non-linear phenomena: Levels of complexity**

When an ODE is expressed by a linear equation of a dependent variable or a derivative of the dependent variable in linear form or both as separate terms, the ODE is linear; otherwise, it is called a non-linear ODE. In particular, ODEs without any products of the dependent variable (either by itself or other functions of itself or its derivatives) can be considered linear. The linearity of independent variables will have no influence over the linearity of the ODE. As an exercise, answer the problem below.

Problem 1.1

Classify the following ODEs as either linear or non-linear.

$$T'' + 2T = 0 \qquad\qquad (1.3)$$

$$T'' + t^3 T = 0 \qquad\qquad (1.4)$$

$$T'' + 2T^2 = 0 \qquad\qquad (1.5)$$

Answer   Equations 1.3 and 1.4 are linear ODEs, while eq. 1.5 is non-linear. In eq. 1.4, even if the second term has the independent variable t raised to the third power, it will not make the ODE non-linear. Classified to be a non-linear ODE, eq. 1.5 has the dependent variable T raised to 2.

Knowing whether an ODE is linear or non-linear is the first step to solving the ODE since the approach to solving them is very different. Non-linear ODEs are much more complex that sometimes approximate solutions are more preferred than exact solutions. Non-linear ODEs represent non-linear phenomena (e.g. forecasting precipitation, complex food pyramids) that are extremely complicated and much more diverse compared with linear phenomena.

Let us take a look at how the first-order term or linear term of the dependent variable in the ODE, its derivative, and its product (non-linear term), model the physical phenomena. As an example, a trigonometric function, $T = \sin \alpha t$, is being substituted to a linear or non-linear term. Here, the wave exhibited by the trigonometric function can represent any cyclic change of a physical phenomenon (e.g. power supply, temperature).

Linear	First-order terms and their derivatives
	$T = \sin \alpha t$
	$T' = \alpha \cos \alpha t = \alpha \sin\left(\alpha t + \frac{\pi}{2}\right)$
	$T'' = -\alpha^2 \sin \alpha t$
Non-linear	Exponent terms
	$T^2 = \sin \alpha t \times \sin \alpha t = \frac{1-\cos(2\alpha t)}{2}$

The resulting wave of the trigonometric function of the independent variable (e.g. t from the example above) are defined by the amplitude (coefficient of the trigonometric function), angular frequency (coefficient of the independent variable), and phase (increments or decrements within the trigonometric function). After substituting a trigonometric function into the terms of a linear ODE, the nature of the wave remains the same with only the amplitude and the phase changing (upper part of 1.4)[20]. However, substituting the function into the terms of a non-linear ODE will produce new trigonometric functions of varying natural frequency[21]. A non-linear system will result in significant changes in the nature of the wave entirely (lower part of Figure 1.4). Complicated and diverse physical phenomena such as chaos and turbulence will be discussed in the latter part of this textbook.

◆ **Homogeneous vs. non-homogeneous ODE**

Another way to classify ODE is in terms of homogeneity. Determining whether an ODE is homogeneous or not is done by checking the terms of the ODE. For example, given an ODE of the form $T'' + p(t)T' + q(t)T = R(t)$, if $R(t)$ is zero, then the ODE is homogeneous. If $R(t)$ is another function of the independent variable or a constant, then the ODE is non-homogeneous[22]. This term physically corresponds to an external force[23].

▷**20**   After differentiation, the sine function becomes a cosine function. This also means a shift in phase by $\pi/2$ in terms of a sine function. Thus, the phase is also changed in the linear term. To visualize this, imagine the amplifier as volume of a musical piece. In this case, the piece's frequency is unchanged.

▷**21**   Taking the same musical piece as an example, the original melody is completely altered.

▷**22**   A common definition stated in most textbooks.

▷**23**   Explained in the later chapters and in Figure 1.5.

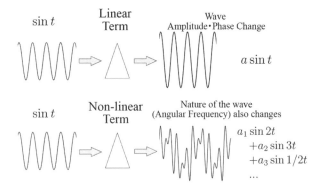

Figure 1.4   Conceptual framework comparing linear and non-linear functions

A more generic definition for homogeneity can be stated. ODEs that have the same order for all variables of the respective terms constituting the equation are considered homogeneous. $R(t) \neq 0$ simply means that there is a term where the dependent variable T is raised to the power of 0 (which is different from the other terms whose order of T is greater than 0).

In applications to physical phenomena, knowing the difference between homogeneous and non-homogeneous ODE can be convenient in determining whether the internal system is experiencing external influence. Moreover, homogeneous ODEs are relatively easier to solve than non-homogeneous ODEs. However, the difference in difficulty to solving homogeneous and non-homogeneous ODEs is not as wide as that of the difference between linear and non-linear ODEs.

2nd order non-linear **homogeneous** ODE $T''T'^2 + 2T^3 = 0$

The example above is a 2nd order ODE because the highest derivative has an order of 2. The equation is non-linear because the total order of the dependent variable is 3 (i.e. greater than 1). Finally, the equation is homogeneous since the sums of the order of the dependent variables for each individual terms (regardless of the order of the derivative) are equal.

2nd order non-linear **non-homogeneous** ODE $T'' + 2T^3 = 0$

In the example above, removing T'^2 from the first term of the previous example will result in a non-homogeneous ODE. The main reason is that the individual sum of the order of dependent variable of the 1st and 2nd terms are no longer equal (i.e. the first term's dependent variable is of order 1, while the second term is of order 3). In investigating physical phenomena, what is more important than deciding homogeneous/non-homogeneous is the non-linearity of the ODE. The ODE above is also non-linear.

1st order **homogeneous** ODE $T' + 2\left(\frac{T}{t}\right)^3 = 0$

In the example above, it appears to be non-homogeneous because the order of the dependent variable T of the second term is more than that of the first term[24]. However, considering the independent variable at the denominator having the same order as the numerator will classify the ODE to be homogeneous since T is a function of t. In the case of 1st order ODEs satisfying this condition, the order of the dependent variable of the terms can be homogeneous depending on the order of the independent variable. In other words, the following should be satisfied,

> For a 1st order ODE $T' = f(T,t)$,
> $f(aT, at) = f(T,t)$
> the form $T' = f(T/t)$ can be fulfilled. (a is a real number)

Try to answer the problem given below.

25 Grouping independent and dependent variables together, $T' = dT/dt$ has a 1st order numerator and a 1st order denominator $(T/t)^3$ has a 3rd order numerator and a 1st order denominator.

Problem 1.2

Classify the following two linear ODEs as either homogeneous or non-homogeneous.

$$T'' + 2T = 0 \tag{1.6}$$

$$T'' + 2T = 3 + t \tag{1.7}$$

Answer Based from the results above, eq. 1.6 is homogeneous while eq. 1.7 is non-homogeneous because of the presence of a constant and independent variable at the right-hand side of the equation (i.e. right-hand side has a zero order of dependent variable). In other words, the order of the dependent variable within the terms are not the same since the right-hand side of eq. 1.7 has an order of 0.

◆ **Superposition principle: Linear homogeneous ordinary differential equation**

In general, for linear homogeneous ODE, the sum of individual particular solutions (with no limit as to the number of summation) and the individual particular solutions are solutions to the equation. This is called the **superposition principle** and is demonstrated by eq. 1.8.

$$T = aT_1 + bT_2 \quad (a, b \text{ are constants})$$
$$f(T) = f(aT_1 + bT_2) = af(T_1) + bf(T_2) \tag{1.8}$$

Even if eq. 1.3 and eq. 1.4 are of 2nd order, the superposition principle is applicable because both are linear homogeneous ODEs. On the contrary, the principle does not apply to eq. 1.5 because it is a non-linear ODE[25].

Even if an ODE is linear, superposition principle cannot be applicable for non-homogeneous ODEs (e.g. eq. 1.7)[26]. However, the principle can still be useful as an intermediate procedure to solving complicated ODEs as will be discussed later. Proof and demonstration of the superposition principle (eq. 1.8) based on problem 1.1 is shown below.

eq. 1.3, $T'' + 2T = 0$:

$$(aT_1'' + bT_2'') + 2(aT_1 + bT_2) = a(T_1'' + 2T_1) + b(T_2'' + 2T_2) = 0$$

25 $(aT_1'' + bT_2'') + 2(aT_1 + bT_2)$
$= a(T_1'' + 2T_1) + b(T_2'' + 2T_2) = 0$
$\tag{1.3}$
$(aT_1'' + bT_2'') + t^3(aT_1 + bT_2)$
$= a(T_1'' + t^3 T_1) + b(T_2'' + t^3 T_2) = 0$
$\tag{1.4}$
$(aT_1'' + bT_2'') + 2(aT_1 + bT_2)^2$
$\neq a(T_1'' + 2T_1^2) + b(T_2'' + 2T_2^2)$
$\tag{1.5}$

26 eq. 1.7 is linear but non-homogeneous, so superposition principle is not applicable.
$(aT_1'' + bT_2'') + 2(aT_1 + bT_2)$
$= 3(a+b) + t(a+b)$
$\neq 3 + t$

eq. 1.4, $T'' + t^3 T = 0$:

$$(aT_1 + bT_2)'' + t^3(aT_1 + bT_2) = a(T_1'' + t^3 T_1) + b(T_2'' + t^3 T_2) = 0$$

eq. 1.5, $T'' + 2T^2 = 0$:

$$(aT_1 + bT_2)'' + 2(aT_1 + bT_2)^2 \neq a(T_1'' + 2T_1^2) + b(T_2'' + 2T_2^2)$$

eq. 1.7, $T'' + 2T = 3 + t$:

$$(aT_1'' + bT_2'') + 2(aT_1 + bT_2) \neq 3(a + b) + t(a + b) \neq 3 + t$$

1.1.4   The first solution: The basic concept

◆ **System as backbone of physical phenomena, physical properties to determine identity**

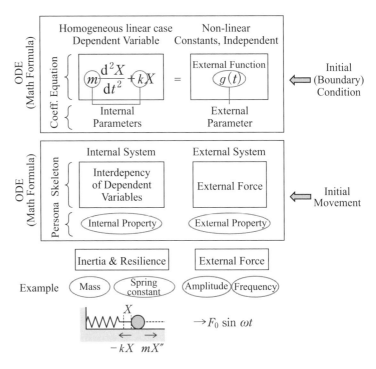

Figure 1.5   Physical phenomena and composition of ODE

The analogy between ODE and physical phenomena is shown in Figure 1.5. An ODE is composed of equations and coefficients. Expressions (or equations) that govern the physical phenomena are collectively called a system[27]. A system is the basic skeleton of the physical phenomena. Examples of systems include machines, electric devices made up of a system of circuits, and the food chain occurring in nature. Inside a system, various exchanges and processes are taking place. An equation representing a

⮌**27**   Returning to Figure 1.1, the system expressed is a tree, and its physical properties are represented by coefficients.

function within a system include "coefficients" which represents **parameters** that describe physical properties influencing the behavior of the system or the physical phenomenon. These parameters can be thought of as a numerical constant that quantifies the specification, performance, or degree of influence of certain parts within a system (recall Figure 1.1)[28]. If a system is the skeleton of a physical phenomenon, the parameters determine the individuality or uniqueness of the phenomenon. The dependent variable of the system varies accordingly with the independent variable following the initial and boundary conditions, and the ODE mathematically representing the system. Thus, solving the ODE is required to describe the changes in the dependent variable.

To understand better the link between mathematical formulae and physical phenomena, it is advisable to separate terms that include dependent variables (left-side terms of Figure 1.5) and terms that don't (functions with only constant terms and independent variable). The isolated terms containing the dependent variables expresses the interdependence of the parts within the system, thus, referring to this group of terms as the internal system. On the other hand, terms that exclude dependent variables can be interpreted as that which comes outside of the system regardless of the condition of the dependent variable. These terms collectively refer to the external system[29]. Considering the motion of a mass connected to a spring and subjected to an external force (as shown at the bottom of Figure 1.5), internal system is comprised of the inertial force proportional to the second order differential of the displacement, which is the dependent variable, and the restoring force of the spring proportional to the displacement. The external forces acting on the spring system such as $F_0 \sin \omega t$ comprise the external system.

◆ Useful analyses charts: Time-series diagram and phase diagram

In order to physically interpret (or analyze) the obtained solution of an ODE, visualizing the solution can be helpful. In this book, visualization (or post-processing) exercises will be conducted. There are two types of plots (common term for the figures or images created for analyses) that will be frequently created throughout this book. The first type is called a **time-series chart** where the horizontal axis and vertical axis are represented by the independent variable (for this case, time) and dependent variable, respectively. If the horizontal axis is represented by spatial coordinates, the chart is referred to as spatial distribution diagram (or chart). The second type of plot has the dependent variable along the vertical axis with another term comprised of a dependent variable (e.g. gradient of the dependent variable) along the horizontal axis. This plot is called a **phase diagram**. The other term can be the derivative of the dependent variable or, in the case of simulatenous ODEs, terms comprised of two or more dependent variables. Phase diagrams contain no information of the independent variable and provides additional perspective and understanding of the physical phenomenon[30].

⇨**28** In Figure 1.5, the specifications of the mass–spring system is represented by the coefficients (parameter).

⇨**29** The distinction between "internal" and "external" systems are introduced in this book to have better correspondence between ODEs and the physical phenomena.

⇨**30** The time-series or spatial-distribution diagram visually describes the solution of the ODE because it plots the functional form of the dependent variable relative to the independent variable. In other words, it represents the behavior of a system (physical phonemena). On the other hand, phase diagram illustrates the relationship between the dependent variable and its derivative. It expresses the interaction of the internal system.

◆ Physical phenomena and ODE introduced in this book

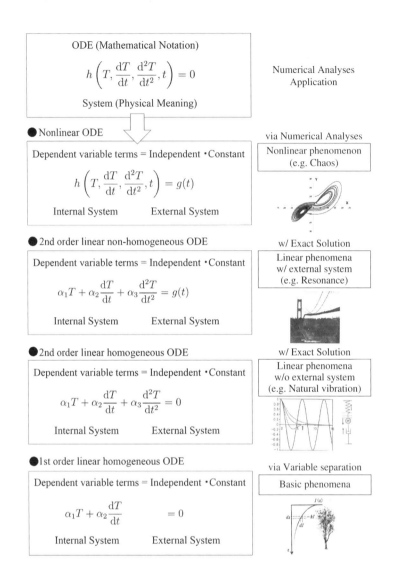

Figure 1.6 Sketch and summary of ODEs and physical phenomena focused in this textbook

Figure 1.6 shows a summary of the contents of this textbook. The concepts introduced is immediately followed by a hands-on exercise to visualize and interpret the solution. It is important for an engineer to understand each term of the differential equation in relation to the physical phenomenon it represents. In 2nd order linear homogeneous ODEs (ch. 3), oscillations will be introduced. An oscillation of an dependent variable can occur only from the interaction of the internal system[31]. When a periodic external force acts as an external system (2nd order non-homogeneous ODE), a surprising phenomenon called resonance may occur (ch. 4). Resonance occurs when the vibration unique to the internal system coincides with the vibration period of the external system. In this phenomenon, the

⟳**31** In the absence of damping, this is called natural vibration. This corresponds to a phenomenon in which a dependent variable is temporally (or spatially) oscillated by the interaction of the internal system, regardless of the absence of external system.

amplitude of the dependent variable's oscillation increases steadily until the system can no longer be sustained and will collapse. Resonance was one of the causes of accidents or engineering disasters in the past but was already investigated extensively in engineering[32]. It is not enough to limit our understanding to linear ODEs. Thus, non-linear ODEs representing non-linear, complex, and diverse behavior are introduced in ch. 6). Regardless of the absence of an external system, non-linear interactions within the system will lead to complex, self-excited, and irregular vibrations called chaos. Nevertheless, the solution of chaotic systems does not deviate far from the system's expected behavior (still satisfying the differential equation). In chaotic systems, slight changes in the initial conditions lead to significant differences in the behavior. When phase diagrams are constructed, a strange pattern called strange attractors may appear. Large strange attractors are also widely used in contemporary art[33].

◆ Law, model, or rule-of-thumb?

A system of equations that is applicable to the widest number of physical phenomena and with the highest reliability (or universality) can be considered a **law**[34]. Newton's laws are typical examples. Although not as universal as a law, a **model** is another system that has a relatively wide range of application and is constructed after clarifying the conditions of the physical phenomenon of interest. An example is the logistic model. A logistic model[35] is a 1st order ODE that predicts the temporal change in the number of organisms. The rate of change (1st order differential) of population N can be expressed by the following ODE (eq. 1.9).

$$\frac{\mathrm{d}N}{\mathrm{d}t} = (\alpha_1 - \alpha_2 N)N \tag{1.9}$$

The rate of change (left-side of eq. 1.9) of population should be proportional to the population growth represented by the term $a_1 N$ but limited to a certain amount by the $a_2 N^2$ if N population becomes too large. Since all terms of the equation contain the dependent variable N, the equation is representative of the internal system (e.g. earth or ant aquarium) and factors within the system define the population N with external factors disregarded. Although the assumption is bold (thus classified as a model rather than a law), it can logically explain variations in population of a large number of the same species. a_1 and a_2 are internal parameters of the system. The uncertainty of these parameters exhibits the lack in universality of the logistic model, since a_1 and a_2 are not physical properties that can be determined a priori and common for all types of populations. In the real world, these internal parameters are dependent on other factors that are changing, such as the change in the control volumne's food supply, the existence of other species, and other environmental factors.

It is also possible to derive a solution empirically. For example, suppose that the change in the number of a certain specie kept in an aquarium is recorded daily. Using excel or any regression software, a curve can be fitted empirically (eq. 1.10), such that the change in population is found to be proportional to the cube of time. The resulting empirical function when plotted together with logistic model will more or less provide a close match.

▷**32** Resonance is often caused by the interaction of natural vibrations of man-made structures (e.g. bridges, houses, warehouses) and periodic natural external forces such as winds and earthquakes.

▷**33** Generally called "fractal art." Apart from its purpose of expressing physical phenomena, phase diagram can also be used as art.

Taken from http://lcl.web5.jp/prog/ sozai/gazou.html

▷**34** There is no way to confirm it strictly "universal." It is better to assume an applicability criteria or condition for every law.

▷**35** Refer to ch. 2.

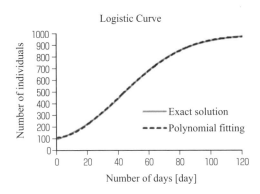

Figure 1.7 Model comparison: Logistic model vs. empirical function

(Figure 1.7).

$$\frac{\partial N}{\partial t} = at^3 + bt^2 + ct + d \tag{1.10}$$

The physical grounds linking the rate of population change with the cube of time as shown by the empirical equation above (eq. 1.10) are thin. An empirical function does not have universal applicability and largely limited to the case being analyzed (recall the case of determining a room's temperature based on a city's power consumption). Its reliability would depend on the amount of sample data used to derive it. Despite this, empirical functions can be the first step to confirming a physical relationship or the design of a model↷**36**. If the functions and the terms of an equation allow physical interpretation (unlike eq. 1.10 where the terms are difficult to physically interpret), it becomes a model. If the versatility of the model is historically recognized over the years with no exceptions, a model may become a law (Figure 1.8).

↷**36** On extremely rare occassions.

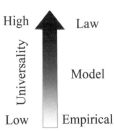

Figure 1.8 Universality of law, model, and empirical function

◆ Determining initial and boundary conditions

In modeling the behavior of a dependent variable, initial and boundary conditions need to be determined in advance. Example would be weather modeling where it is necessary to specify meteorological conditions at the initial time in order to know the next time's condition. In modeling species inside a control volume, it is necessary to know the influx at the walls or at the opening of the control volume (boundary condition).

It has been previously stated that n-initial conditions and boundary conditions are necessary to obtain an exact solution for n-order of ODEs. Let us see in more detail how to reasonably determine those conditions from physical phenomena of interest by taking a second order ODE as an example.

When the independent variable is time, the dependent variable and its derivatives (e.g. velocity, acceleration) during the initial time of the target physical phenomenon are required[37]. Is it possible to set time conditions other than at initial $t = 0$? The answer is yes. Cases where the value of a dependent variable is known at an arbitrary elapsed time $t = t_I$ is quite common and can be considered mathematically. However, as will be encountered in the later chapters, applying this inside a numerical solution can be more problematic than the exact solution. This is because the physical state at the time t_1 arbitrarily cut out of the numerical solution is merely the result of integration in the time direction from the initial condition. In investigating physical phenomenon, the stage when the initial condition is given determines a unique behavior of the dependent variable from any stage of initial time from negative infinity to positive infinity. In engineering, the physically meaningful time can be the initial time[38][39]. Determining a physically meaningful initial condition at a given time and space can be quite challenging.

Next, consider the boundary condition when the independent variable is a spatial coordinate. Similar with time coordinate, when considering the spread of space from one boundary to infinity, two dependent variables and its differential value (spatial gradient) at the end points are required as boundary conditions. If the target space has two end points, like a finite bar, the dependent variable value at one end-point and the dependent variable value or its differential value (spatial gradient) at another or the same end-point (total of two boundary conditions) is necessary.

[37] Setting the intial time is also part of the problem.
- Physically clear definition of "initial" time such as the launch time of a rocket.
- Motion of stars where the true initial state is unknown but defined by the time when observation is initiated (artificial time).

[38] In numerical analyses, trial-and-error of the intial condition is possible to find a desired behavior of the system.

[39] This also comes to mind causal or scientific determinism which states that for atoms with known location and momentum, its past and future is already determined. In the history of science, this is referred to as "Laplace's demon."

◆ Dirichlet (strong) vs. Neumann (weak) conditions

Table 1.3 Influence of the internal parameter to the solution

Dirichlet Condition	Neumann Condition
Known dependent variable value	Known derivative
Strong constraint condition (1 required)	Weak constraint condition
Ex. Temperature distribution along a bar (Temperature T)	Ex. Heat flux on the wall (K: Diffusion coefficient)
$T_{x=0} = 300$ $x = 0$	$K\left(\dfrac{dT}{dx}\right)_{x=0} = 0$ $x = 0$
Ex. Deflection of beam (Vertical displacement Y)	Ex. Deflection angle of the wall
$Y_{x=0} = 0$ $x = 0$	$\left(\dfrac{dY}{dx}\right)_{x=0} = 0$ $x = 0$

In modeling, initial (boundary) conditions can be considered to be

Dirichlet or **Neumann**. Dirichlet condition is when the known initial (boundary) condition is a constant value of the dependent variable, while Neumann condition is when the known condition is a constant value of the first derivative of the dependent variable with respect to an independent variable (Table 1.3). Dirichlet condition is considered to be a strong constraint since a value of the dependent variable is already assumed or known↩**40**. On the other hand, Neumann condition is considered a weak constraint because only the derivatives are known. While the purpose of a Dirichlet condition is obvious, what kind of physical phenomena is represented by a Neumann condition?

When the independent variable is time and displacement (distance from one point of origin to another) is the dependent variable, Neumann condition means a 1st order derivative of displacement with respect to time (or speed) is known. Furthermore, it could also be called Neumann condition if the second derivative (acceleration) is known. If the independent variable is space (x as shown in Table 1.3), 1st order differential refers to gradient/slope (1st derivative) or unevenness (2nd derivative). Various phenomena can be represented by the Neumann condition. An example would be heat flux by diffusion. Heat flux refers to the amount of heat passing through a substance per unit time and unit cross-sectional area by molecular diffusion. It is known to be proportional to the temperature T gradient of the substance. If the heat flux (or mass flux) at an end point is known, the corresponding known spatial gradient is called a Neumann boundary condition. This boundary could be likened to an insulated covering of a material at one end which allows no flux (i.e. gradient or 1st derivative of T is 0 at the wall).

Another example would be an overhang (e.g. diving board for swimmers, terrace). If horizontal distance is the independent variable and the displacement of the edge of the overhang (deflection, Y) is the dependent variable, the 1st order differential is the deflection angle (dT/dx) and the 2nd order differential is the curvature. The overhang is fixed to the wall (called the fixed end) which means that at the position closest to the wall, vertical displacement and deflection angles are 0. This case involves a combination of Dirichlet and Neumann conditions.

Dirichlet and Neumann conditions can be combined but at least one Dirichlet condition should be known if an exact solution is needed. The provision of the dependent variable value by the Dirichlet condition is a stronger condition, and it is impossible to unambiguously determine an exact solution of the ordinary differential equation merely by specifying the differential value using the Neumann condition.

n initial conditions and boundary conditions are necessary to obtain an exact solution of an nth oder ODE, it can never be more or less than the order of the ODE. When solving ODEs approximately using a computer, mistakes to implement a physically realistic initial (boundary) conditions is quite common. Thus, careful treatment is necessary.

◆ Variety of solutions: Its advantages and disadvantages

While "general," "particular," and "exact" solutions, are similar except for the treatment of the constants of integration, another approach to solving

↩**40** From Table 1.3, the Dirichlet condition defines the value of the dependent variable for a certain value of the independent variable. This allows the determination of the constant of integration. Neumann condition only defines the value of the derivative of the dependent variable; thus, the constant of integration remains arbitrary.

ODEs is through numerical methods. Aside from deriving the exact solution to an actual physical problem, numerical methods can be used to solve an ODE. The main differences between both solutions are summarized in Table 1.4. Conventional methods have been established to determine the exact solution of various ODEs. The variable separation method, which is a rudimentary solution, is based on the condition that the dependent variable and the independent variable in the ODE are separable into both sides of the equation (left and right of =), and integration is performed on each side. Although beyond the scope of this textbook, integral transforms such as Fourier transform is useful to reduce differential equations to algebraic equations. Numerical solutions provide approximations of the exact solution. Numerical methods are more common in practice because of its very wide applicability. Numerical methods is considered to be only an approximation of the exact solution because it approximate the derivatives using algebraic equation based on simple arithmetics.

The exact solution means that it is a functional form of the dependent variable exactly obtained from the ODE. Exact solutions have no room for error. Furthermore, the overall behavior of the system can easily be understood from the functional form. Unfortunately, the conventional methods to arriving at an exact solution have limited applicability and usually not applicable for non-linear ODEs. On the other hand, numerical methods are, in principle, generally applicable to most ODEs.

Table 1.4 Exact solution vs. approximate solution

Solution type	Exact solution	Numerical solution
Solution name	Variable Separation Method 2nd Linear ODE solution Integral transform (Fourier, Laplace)	Runge–Kutta Method Euler Method
Applicability	Narrow (Almost for linear)	Wide
Error	None	Yes

In addition, advanced computational softwares have been developed with high speed and accuracy. Despite this, the limitation of numerical methods is that it will always remain an approximation and inspection is frequently needed to minimize its errors. If numerical methods are not set-up properly, the numerical solution may behave completely different from the exact solution. In addition, since the numerical solution does not calculate the functional form of the dependent variable unlike the exact solution, it only outputs the dependent variable values corresponding to the specific independent variable values further dictated by resolution or spacing of independent variables. The behavior of the physical phenomena is not known until figures (e.g. time-series or phase diagrams) or post-processing is conducted. More on this will be discussed in the later chapters.

1.2   Very basic Python tutorial

In this textbook, Python Language will be used for the PC exercises. In this section, a brief tutorial is discussed. For details on how to download and install Python for your system, refer to https://wiki.python.org/moin/BeginnersGuide/Download.

Once you have successfully installed Python in your system, you can successfully type the command "python" without error in the command prompt (Windows), or terminal (MacOS or Linux). Begin by entering "python" at the command line. This will allow you to operate Python in interactive mode which is useful for testing modules or commands.

While in interactive mode, enter the command

```
1   print("Hello␣World!")
```

"Hello World!" will be displayed. This shows how simple it is to actually print a command.

How are variables created and how are arrays stored? In the same interactive mode, enter the following in sequence.

```
1   A = 1.
2   B = 2.
3   C = 3.
4   D = A+B+C
5   print(A+B+C)
6   print(D)
```

Upon completing the sequence of images, "6.0" will appear first and another "6.0" will appear second. The first result shows the value of the sum of A, B, and C. The second result shows the value of D which stores the sum of A, B, and C.

To create an array, the following commands may be tested in interactive mode.

```
1   A = [1.,2.,3.,4.]
2   B = [1.,2.,3.,4.]
3   C = A+B
4   print(C)
5   print(A+B)
```

This time, the result shows [1.0, 2.0, 3.0, 4.0, 1.0, 2.0, 3.0, 4.0]. This means that the lists are simply appended such that the list of values for B are appended to A. This is also shown twice since we printed out the value of C which corresponds to A+B.

If we then type, D = A*B, an error will appear that states "TypeError:

can't multiply sequence by non-int of type "list"." Inspecting and correcting this error is commonly called "debugging." Can you debug the error? The error means that the variables created so far are of type "list." In Python, this is not exactly a numerical array. For numerical arrays, we may begin relying on a module called "NumPy."

To demonstrate this very useful module, type the following commands in sequence in interactive mode.

```
1  import numpy as np
2  A = np.array([1.,2.,3.,4.])
3  B = np.array([1.,2.,3.,4.])
4  C = A+B
5  D = A*B
6  print(C)
7  print(D)
```

This will result in arrays of the same size as A corresponding to element-wise sum and multiplication of A and B arrays: [2. 4. 6. 8.], [1. 4. 9. 16.]. Throughout this textbook, these arrays which can be called "NumPy arrays" will be widely used. For more information about NumPy, refer to `https://numpy.org/`.

One of the advantage of Python is its capability to use advance modules for plotting images or figures. For example, one may plot A against B using the following sequence of commands continuing with the previous commands.

```
1  import matplotlib.pyplot as plt
2  plt.scatter(C,D)
3  plt.show()
```

The following figure is produced.

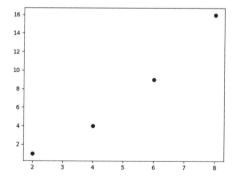

◆ ◆ ◆ ◆ ◆ ◆ **Problems for Exercise** ◆ ◆ ◆ ◆ ◆ ◆

1.1 Classify the following equations as either ordinary differential equations or partial differential equations.

1 $T_t = e^{-t}T_{tt} + \sin t$

2 $T_{tt} = e^{-t}T_{xx} + \sin t$

3 $TT_{xx} + T_x = 3$

4 $T_{xx} + xT = 0$

5 $xT_x + tT_t + T^2 = 0$

6 $\begin{cases} T_{xx} + C_x + CT = 1 \\ C_{xx} + C = \sin x \end{cases}$

7 $\begin{cases} T_{xx} + C_t + CT = 1 \\ C_{xx} + C = \sin t \end{cases}$

1.2   Answer the following questions regarding the ODEs determined from the previous problem.

1 How many initial conditions or boundary conditions are needed to find an exact solution?

2 Is the ODE linear or non-linear?

3 Is the ODE homogeneous or non-homogeneous?

2 1st order ODE

Various physical phenomena tackled in engineering that are seemingly different such as a free-falling mass, organism's microbial growth, electric circuits, forest-intake of solar radiation, and box model, can be expressed by simple 1st order ordinary differential equations (ODE). The widely used general solution of 1st order ODE which is commonly in the form of exponential functions can be derived using the variable separation method, which will then be used to acquire the exact solution[41]. The discussion is to be followed by a PC exercise to interpret the phenomena being considered.

[41] In addition to the variable separation method, other methods are available for obtaining the exact solution which either reduces to the same method or other complex methods. This textbook focuses mainly on the variable separation method.

2.1 Variable separation method

A 1st order ODE which can be represented by the product of two functions where one consists only of the dependent variable and another only of the independent variable is called a **separable ODE**. An example is an ODE of the form shown below (eq. 2.1).

$$\frac{dT}{dt} = g_1(T)g_2(t) \tag{2.1}$$

Separating the dependent terms and independent terms to either sides of the equation[42] (i.e. transposing the dependent term to the side of the derivative terms), and integrating both sides (eq. 2.2) leads to a general solution. Exact solutions are obtained once the constants of integration are substituted with values corresponding to known initial (boundary) conditions.

[42] Multiply $dt/g_1(T)$ to both sides.

$$\int \frac{1}{g_1(T)} dT = \int g_2(t) dt \tag{2.2}$$

2.2 Variable separation of 1st derivatives of linear expresssion

2.2.1 Exact solution and physical meaning

Eq. 2.1 can be written in the form as shown below (eq. 2.3) where the time (t) rate of change of a dependent variable (T) is equal to a linear expression

◌43 Recalling Figure 1.5, an internal system is a system comprised mainly of terms containing the dependent variable and the interactions among them. In eq. 2.3, the terms dT/dt and aT include the dependent variable. An external system is a system not directly related with the dependent variable. In eq. 2.3, it is represented by the constant term b.

of the dependent variable with constant coefficients (a,b). The coefficients are the parameters that define the uniqueness of the physical phenomenon. By the way, can you recall which terms represent the internal system? a is an internal system parameter which defines the time rate of change of the dependent variable, while b is a constant external system parameter[◌43]. Since the mathematical expression expresses a physical phenomenon, the physical units in each term of the expression must be consistent. For example, given that the unit of time is seconds, and T is K (which stands for Kelvin), a must have a unit of $1/s$, and b must be a parameter having a physical unit of K/s.

$$\frac{dT}{dt} = aT + b \tag{2.3}$$

Given an initial condition $T_{t=0} = T_0$, an exact solution for eq. 2.3 can be acquired using the following procedure.

$$\int \frac{1}{aT+b} dT = \int dt$$

$$\frac{1}{a} \ln|aT+b| + C_1 = t$$

$$\ln|aT+b| + C_2 = at$$

$$aT + b = C \exp(at)$$

$$aT_0 + b = C \quad \text{(substituting the initial condition)}$$

$$aT + b = (aT_0 + b) \exp(at)$$

◌44 Feedback means to return information or output from the system to the system input in some form. Applying eq. 2.3 to global warming for example, a positive feedback ($a > 0$) means that if there is an increase of $1°$C to a $20°$C leading to $21°$C, the increased temperature further promotes a higher temperature rise. In the case of negative feedback system ($a < 0$), the opposite will be experienced, such that higher temperature leads to higher temperature decrease, which means a suppression of temperature rise.

Exact solution: $T = \left(T_0 + \dfrac{b}{a}\right) \exp(at) - \dfrac{b}{a}$ $\tag{2.4}$

Try to consider the physical meaning of the solution. First, look at how the independent variable influences the dependent variable. Temperature exponentially varies with time ($\exp(at)$). Furthermore, the influence of time is amplified by the given initial condition, T_0, and the ratio of parameters b (external) and a (internal). The second term, $-b/a$, serves as the intercept of the equation which somehow represents the scale of the control volume of the dependent variable being considered. Finally, notice how the sign of the internal parameter a influences the behavior of the phenomenon. Depending on the sign, a causes either a positive or negative feedback to the solution (see Table 2.1)

◌45 If a is 0, eq. 2.3 may be used instead of eq. 2.4
$dT/dt = b$
$T = bt + T_0$

When $a > 0$, the dependent variable T has an increasing time-rate of change dT/dt, further promoting its increase with time. Such a system is called a positive feedback system[◌44]. On the contrary, when $a < 0$, the dependent variable T has a decreasing time-rate of change dT/dt, further promoting its decrease with time. This system is a negative feedback system. Below, let us examine various physical phenomena which can be rep-

Table 2.1 Influence of the internal parameter a to the solution

$a > 0$ (a is positive): Positive feedback	T increases exponentially, T diverges as $t \to \infty$
$a = 0$ (a is zero): No feedback (eq. 2.3)	Constant rate of T change (i.e. monotonous increase with time)[◌45]
$a < 0$ (a is negative): Negative feedback	Exponential attenuation of T, T converges to $-\frac{b}{a}$ as $t \to \infty$

resented by eq. 2.3 and its solution (eq. 2.4).

2.2.2 Free-falling body

Let us predict the time rate of change of the velocity v of a free-falling body with a given mass m (Figure 2.1)[46]. The body encounters air resistance as it falls which is proportional to its speed by c (proportional constant). Its initial velocity is zero (i.e. $v_{t=0} = 0$). It is obvious that v is the dependent variable and time t is the independent variable. Using Newton's law of motion that forces acting on a mass point balance at any certain inertial reference, we can express the phenomenon using ODE (eq. 2.5). The forces acting are the inertial force, the air resistance force, and gravitational force. The inertial and air resistance forces comprise the internal system while the gravitational force comprises the external system.

[46] In the case of a small mass-point such as a rain droplet, the air resistance can be assumed to be proportional to the speed of fall. However, the air resistance can also be proportional to the square of the speed such as a person riding a bicycle. In this case, the problem becomes non-linear and cannot be solved by the method introduced in this chapter.

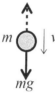

Figure 2.1   Free-falling mass point

$$m\frac{\mathrm{d}v}{\mathrm{d}t} = -cv + mg \qquad (2.5)$$

Maintaining the same physical units for each term, the unit of the resistance coefficient c is kg/s, or mass unit per time[47]. Eq. 2.5 can be simplified further by dividing both sides of the equation by mass (kinematic form). This reduces the number of parameters into two, namely; an internal parameter $a_1 = c/m$[48] and the external parameter g. The resulting equation is shown in eq. 2.6.

[47] Substituting the units of mass m [kg], velocity v [m/s], gravitational acceleration g [m/s^2], and resistance coefficient c [kg/s] to the right-hand side of eq. 2.5 results in a consistent unit [kg·m/s^2].

$$\frac{\mathrm{d}v}{\mathrm{d}t} = -a_1 v + g \qquad \text{where } a_1 = \frac{c}{m} \qquad (2.6)$$

[48] Note that this parameter, representing a physical property, will always have a positive value.

From eq. 2.5, two internal parameters can be seen, m and c. However, by rearranging eq. 2.6, the intrinsically important internal parameter is only one which is represented by the ratio of c and m. Here, it is obvious that for any two objects of different c and m but of equal a_1, the behavior of fall will be exactly the same. By analogy with eq. 2.3 and its exact solution (eq. 2.4), an exact solution to the current problem can be written by eq. 2.7.

$$v = \left(v_0 - \frac{g}{a_1}\right)\exp(-a_1 t) + \frac{g}{a_1}$$

$$= \left(v_0 - \frac{mg}{c}\right)\exp\left(-\frac{c}{m}t\right) + \frac{mg}{c} \qquad (2.7)$$

▷**49** Comparing the coefficients of the dependent variable between eq. 2.6 and eq. 2.3, negative feedback system can be inferred for the free-falling body. Oftentimes, comparing the ODE of the system with eq. 2.3 will suggest the behavior of the system.

Here, since the internal parameter $(-a_1)$ will always result in a negative value, it will cause a negative feedback▷**49**. At an initial time given zero velocity, the air resistance of the mass point is also zero. However as the mass point begins to fall at a certain velocity due to gravity, the air resistance correspondingly increases leading to a decrease in the mass point's acceleration. Thus, this behavior suggests a negative feedback. After a sufficient time of fall ($t \to \infty$), the acceleration falls to 0 and the speed becomes constant resulting in a state when gravity and air resistance are balanced. This leads to a state of **terminal velocity** which is equivalent to $v_{t=\infty} = g/a_1 = mg/c$.

2.2.3   Self-propagating/self-destruction of micro-organisms in a container

▷**50** In this phenomenon, problem description and analysis by ODEs is effective even for problems that do not seem to be in the field of physics or engineering, such as changes in the number of organisms. This type of example belongs to an important field of biology called "mathematical ecology."

In the next example, there are two species of microorganisms cultured separately in a sealed test tube (Figure 2.2). One kind is optimistic such that 5% of the number of the existing daily population in the test tube self-replicate by cell division. The other kind is pessimistic such that 5% of the number of their existing daily populations are self-destructing/self-dissolving. Both species are strictly controlled in separately sealed tubes. In addition, factors such as food in-take and living conditions do not adversely affect their reproduction or destruction. However, aside from the processes occurring within each species, 100 microorganisms of the same species are injected into the test tube every day▷**50**.

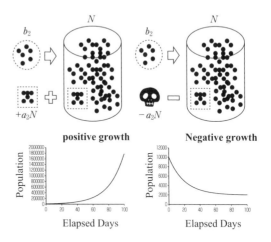

Figure 2.2   Optimistic species (L) vs. pessimistic species (R)

$$\text{Optimistic species} \quad \frac{\mathrm{d}N}{\mathrm{d}t} = +a_2N + b_2, \tag{2.8a}$$

$$\text{Pessimistic species} \quad \frac{\mathrm{d}N}{\mathrm{d}t} = -a_2N + b_2. \tag{2.8b}$$

The rate of change of both optimistic and pessimistic species are shown by the ODEs eq. 2.8. The microorganism's population N (count) is the dependent variable and time t (day) is the independent variable. Optimistic species have a daily increase in population equal to 5% of the number of individuals per day. On the other hand, pessimistic species have a daily decrease in population equal to 5% of the number of individuals per day. Both of these occurrences happen within the internal system and can be expressed by the product of the internal parameter a_2 (reproduction rate). The only differences between the two types are the positive and negative signs acting on the internal parameters. Here, the physical unit of the internal parameter a_2 is 1/day. Based on the given phenomenon, a_2 is equal to the following,

$$a_2 = 0.05 \tag{2.9}$$

In addition, 100 microorganisms are supplied daily. This condition belongs to the external system and is expressed by b_2 with a physical unit of, microorganisms/day,

$$b_2 = 100 \tag{2.10}$$

Given the initial condition that $N_{t=0} = N_0$, eq. 2.4 can be written as,

$$\text{Optimistic species} \quad N = \left(N_0 + \frac{b_2}{a_2}\right)\exp(a_2 t) - \frac{b_2}{a_2} \tag{2.11a}$$

$$\text{Pessimistic species} \quad N = \left(N_0 - \frac{b_2}{a_2}\right)\exp(-a_2 t) + \frac{b_2}{a_2} \tag{2.11b}$$

For the optimistic species, a positive feedback system is formed with the population increasing exponentially to infinity with time[51]. On the other hand, the pessimistic species will encounter a negative feedback and stabilize to a population count of $b_2/a_2 = 2000$.

This example highlights the difference between a "law" and a "model." The solution of eq. 2.5 (eq. 2.7) derived from Newton's law strictly expresses the continuous temporal behavior of the mass point. For this example, the microorganisms are uniquely representing different species (eq. 2.8) and their population can never become negative (eq. 2.11). The limitations of the exact solution only to certain conditions (less applicability) makes it a "model" and not a "law."[52]

2.2.4 Electric circuit composed of resistor and capacitor: RC circuit

Let us consider an example of an electric system (Figure 2.3). An electric circuit (specifically an RC circuit) in which a power supply of constant DC voltage, E [units: V][53], a resistor providing electric resistance R [units: Ω=V/A], and a capacitor with a capacitance C [As/V] are connected in series. Let us predict the time rate of change of the current I [A] flowing in the circuit. The dependent variable is I and the independent variable is time

⮫**51** Species with strong fertility levels may manifest actual exponential population increase. By the way, the world's total human population is also increasing exponentially at the time of this book's writing.

⮫**52** The number of living organisms is always a positive integer. Actual increase (self-propagation) or decrease (self-destruction) is either continuous or discontinuous. Therefore, an actual system cannot strictly obey only one of either equations of eq. 2.8. Rather, these equations can only be "models" where number of individuals, growth rate, and death rate are assumed to be continuous for the purpose of physical interpretation. The solution N (eq. 2.11) may result in a decimal value, but the actual number of individuals can be regarded as a rounded-off value of the calculated N.

⮫**53** Volts V in SI units is equivalent to kg m^2 s^{-3} A^{-1}, which confirms the consistency in physical dimensions of the system.

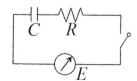

Figure 2.3 RC Circuit

◌**54** Kirchhoff's voltage states, "In any closed loop network,
the total voltage around the loop is
equal to the sum of all the voltage
drops within the same loop."

t. From Kirchhoff's voltage law◌**54**, we can write an ODE for the electric
circuit. First let us define the voltage drops in terms of the current. From
Ohm's law, the voltage drop V_R of the resistance is given by the following
equation,

$$V_R = RI. \tag{2.12}$$

The voltage drop V_C by the capacitor is obtained by dividing the electric
charge amount Q by the capacitance C. Here, the electric charge is equivalent to the time integral ($\int I \mathrm{d}t$) of the current.

$$V_c = \frac{Q}{C} = \frac{\int I \mathrm{d}t}{C}. \tag{2.13}$$

From Kirchhoff's voltage law, the sum of the voltage drops should be equivalent to the supply voltage E. This is represented by the equation,

$$RI + \frac{\int I \mathrm{d}t}{C} = E. \tag{2.14}$$

The left-hand side of eq. 2.14 represents the internal system since each
term contains the dependent variable I. The power supply on the right-hand
side is the action of the external system. When both sides of eq. 2.14 are
differentiated with respect to time, since the DC voltage E remains constant

◌**55** In the case of AC voltage,
the voltage changes periodically
and, thus, acts as an external system. Such system is discussed in
ch. 4.

throughout◌**55**, an ODE without the external system can be obtained.

$$\frac{\mathrm{d}I}{\mathrm{d}t} = -\frac{1}{RC}I \tag{2.15}$$

Assuming that the amount of charge of the capacitor is zero when the circuit
is turned off, the initial current ($I_{t=0} = I_0$) becomes E/R based on eq. 2.14
at the instant when the circuit is switched on. From the above, the exact
solution of eq. 2.15 can be represented by eq. 2.16.

$$I = I_0 \exp\left(-\frac{1}{RC}t\right) = \frac{E}{R}\exp\left(-\frac{1}{RC}t\right) \tag{2.16}$$

From the understanding we obtained in the previous examples, it can immediately be said that the solution is a negative feedback system with the
current eventually converging to zero as time approaches to infinity.

2.2.5 Decay of radioactive elements

◌**56** Given that N is population
density, its dimension is population
per m^3. Since time is in seconds
[s], the units of natural rate of decay will be 1/s.

Another example would be predicting the amount N (dependent variable)
of radioactive elements at a certain time t (independent variable) while experiencing natural decay (Figure 2.4). For every time interval, $\lambda \times 100[\%]$
of the elements naturally decay◌**56**. Ignoring external influence (i.e. absence of external system), the behavior resembles the earlier example of the
pessimistic species in the test tube. An ODE can be written as,

$$\frac{\mathrm{d}N}{\mathrm{d}t} = -\lambda N \tag{2.17}$$

Given the initial condition of $N_{t=0} = N_0$, the exact solution is

$$N = N_0 \exp(-\lambda t). \tag{2.18}$$

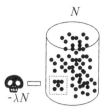

Figure 2.4 Natural decay of radioactive elements

Furthermore, the time required for the number of radioactive elements to reach half of the initial state is defined as half-life $t_{1/2}$ [57] and can be defined by using the exact solution (eq. 2.18).

$$t_{1/2} = \frac{\ln(N_0/N)}{\lambda} = \frac{\ln 2}{\lambda}. \tag{2.19}$$

The half-life (eq. 2.19) depends only on the internal parameter λ and does not depend on the initial state at all. Since λ is specific to a radioactive element, it is possible to identify the radioactive element from half-life measurements [58].

2.2.6 Attenuation of solar radiation (Beer's law)

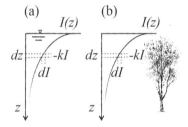

Figure 2.5 Attenuation of solar radiation in water or forest

Let us look at a phenomenon where solar radiation penetrates through the depth of a water body, gradually attenuating and dissipating with depth from the surface of the water (Figure 2.5a). Assuming the turbidity of a water body (e.g. lake) is uniform, we can derive an ODE that can determine the amount of solar radiation for each depth of the water. Uniform turbidity means that the amount of particles (e.g. pollutant, scalar) are equivalent for every location in the water body. Solar radiation is absorbed by the particles such that the remaining solar radiation is lesser with depth. Solar radiation I, the dependent variable, varies therefore with depth z (independent variable) from the water surface [59]. The change in solar radiation per unit depth can be expressed by the product of solar radiation at the depth of water and an attenuation coefficient (κ), an internal parameter representing turbidity or the amount of scalar concentration [60]. The stronger the turbidity, the higher the attentuation coefficient. Assuming uniform turbidity, attenuation coefficient can be assumed constant with depth. If turbidity

> **57** In 2011, when the Japanese version of this book was written, a radioactive leak from the Fukushima power plant occurred due to the tsunami of the Great Tohoku Earthquake in Japan. Various radioactive elements were detected. The media reported its effect to the environment and humans exposed. The time-scale of influence is based on half-life estimates.

> **58** A radioactive element having a long half-life takes a long time to decay and may linger in the atmosphere for a long time.

> **59** Since solar radiation is energy per unit horizontal area per unit time, its units is [W/m^2]. On a peak summer day, it is around 800–1000 W/m^2 in Japan.

> **60** Considering the physical dimensions of eq. 2.20, the units of attenuation coefficient is [1/m].

varies with water depth, the attenuation coefficient becomes a function of the independent variable (z)[61]. The ODE representation is

61 Even if the attenuation coefficient depends on z, variable separation method can still be used to derive the exact solution.

$$\frac{\mathrm{d}I}{\mathrm{d}z} = -\kappa I. \tag{2.20}$$

An exact solution can be obtained by setting $I_{z=0} = I_0$ as the boundary condition of the water surface. The exact solution (eq. 2.21) shows that the solar radiation attenuates exponentially with depth.

$$I = I_0 \exp\{-\kappa z\} \tag{2.21}$$

62 If the solar radiation (I) can be measured at a certain depth z of a lake or sea and at the ground level (I_0), it is possible through eq. 2.21 to obtain an index representing the attenuation coefficient (i.e. turbidity). Similarly, if we can measure solar radiation on an open field (I_0) and assume it to be the solar radiation above the forest, the solar radiation within the forest (I), and the average forest height, the attenuation coefficient (i.e. leaf volume density) may be estimated. It is difficult to estimate individual leaves but this parameter is useful in the field of forestry. The method of estimating the internal parameter from the observed value through ODE is called parameter identification or more generally, an inverse problem.

This example is known as Beer's law[62]. It also describes the interception or reduction of shortwave radiation as it penetrates from a forest crown to the ground (Figure 2.5b). In this case, if an equation to estimate the attenuation coefficients as a function of leaf volume density, leaf slope degree, or leaf type are available, a more reliable model can be constructed. Despite the simplified assumption in Beer's law (eq. 2.21) to set the attenuation coefficient as constant, it has been widely used; thus, it is known to be a law. Furthermore, the system captures the essence of the phenomenon, even if the internal parameter (κ) widely varies.

2.2.7 Balance of physical quantity in space: Box model

The flow of money (i.e. savings) in your bank account can be displayed by sequences of income and expenditures (account statement). Similarly, if you want to predict the temporal change of a certain physical quantity within a specified space, the balance of its physical quantity, its production and reduction in a given space, can be represented by a mathematical expression. A mathematical expression of such physical phenomena is generally called a box model. At any time, a physical quantity flows inside and

63 The unit of flux is a product of the units of the physical quantity of interest and the units of speed (C m/s); and the unit of source and sink is the physical quantity of interest per unit time (C/s).

outside the surface boundary of the inspection space. Flux[63] is a measure of the flow in terms of the rate of increase/decrease of amount per unit area into the volume of interest. Generation and destruction within the control volume per unit time can also be represented by "sources" and "sinks," respectively.

Let us examine a room. Let the volume and the surface area of the boundary be denoted by V [m^3] and S [m^2], respectively (Figure 2.6). Consider the CO_2 concentration C [g] balance in the room (i.e. box model). The source and sink terms are denoted by C_{s_+} and C_{s_-} [g/s], respectively[64].

64 For example, the amount of CO_2 released by humans during respiration per unit time may be considered as source (generated amount). The amount of CO_2 absorbed by plants for photosynthesis may be considered as sink (lost amount).

The flux of concentration inwards and outwards of the room are denoted by $C_{f_{in}}$ and $C_{f_{out}}$ [g·m/s], respectively. The rate of increase or decrease of C per unit time in space (i.e. volume of the room) is equal to the sum of the contribution of the fluxes, sources, and sinks. The following balance formula can be obtained.

$$V \frac{\mathrm{d}C}{\mathrm{d}t} = S \times (C_{f_{in}} - C_{f_{out}}) + V \times (C_{s_+} - C_{s_-}) \tag{2.22}$$

From eq. 2.22, it can be seen that the box model is a 1st order ODE. The equation can be extended to many applications, such as the average

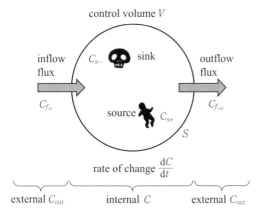

$$\underbrace{}_{\text{external } C_{\text{out}}}\quad\underbrace{}_{\text{internal } C}\quad\underbrace{}_{\text{external } C_{\text{out}}}$$

Figure 2.6 Mass balance of physical quantities in a control volume

substance concentration over a water body[65], prediction of time-varying global average temperature and CO_2 concentrations[66], and so on.

Considering a case where sources and sinks of CO_2 are absent in the room and only its fluxes are present, the difference in the inflow and outflow fluxes will result in a net flux where the sign dictates the direction of net flux (i.e. either inwards or outwards of the control volume). If the outdoor's CO_2 concentration C_{out} is lower than that of the room's, there is no inflow flux and, instead, the outflow flux can be estimated as follows,

$$C_{f_{\text{out}}} = K(C - C_{\text{out}}). \tag{2.23}$$

Here, K [m/s] is a coefficient that represents the proportion of the CO_2 concentration between indoor and outdoor, also called mass exchange coefficient[67]. Given the assumptions above, an ODE to predict the CO_2 concentration of the room can be estimated by substituting eq. 2.23 into eq. 2.22,

$$\frac{dC}{dt} = -\frac{S}{V}K(C - C_{\text{out}}). \tag{2.24}$$

C_{out} is the concentration of the external system. Assuming that the value of C_{out} is constant and the initial CO_2 concentration in the room is C_0, the solution of the ODE becomes[68],

$$C = (C_0 - C_{\text{out}})\exp\left(-K\frac{S}{V}t\right) + C_{\text{out}} \tag{2.25}$$

From eq. 2.25, it can be seen that the CO_2 concentration of the room gradually approaches the outdoor concentration after a certain time, thus the box model here is a negative feedback system. It can also be found that the attenuation rate is proportional to the mass exchange rate and the ratio of the surface area of the boundary and the control volume (i.e. room's or box's volume)[69].

65 Using lakes and ponds as control volumes, quantity of substances flowing in and out of the system, and the generation and depletion of chemicals and its reaction can be estimated.

66 Using the earth's atmosphere as a control volume, quantity of substances flowing in and out of the atmosphere from the ground (cities, plants, or bare land) and the sea surface can be estimated.

67 Can be thought of as ventilation through walls or windows.

68 The solution (eq. 2.3) for the ODE (eq. 2.4).

69 Applying eq. 2.25, let us compare the ventilation efficiency of a small cubic room and a large cubic room. The initial concentration C_0, external concentration C_{out}, mass exchange coefficient K are assumed equal for both rooms. Since the surface area (S) and room volume (V) is proportional to twice and thrice the room's side length, respectively, the surface-volume ratio (S/V) is inversely proportional to the size of the room. Therefore, the larger the room size, the longer the ventilation time.

2.3   1st order non-linear ODE solvable by variable separation method

What if the internal parameter is a factor of the dependent variable? To picture this situation, imagine that instead of having a constant internal parameter a in eq. 2.3, a is assumed to be a function of T. For this case, eq. 2.3 becomes

$$\frac{\mathrm{d}T}{\mathrm{d}t} = a(T)T + b. \qquad (2.26)$$

As already discussed in the first lecture, eq. 2.26 is a non-linear ODE and the situation of the system becomes more complex. However, if the external parameter b is zero, variable separation method can still be applicable to easily derive an exact solution. In some cases, even if the external system is not 0 in a non-linear equation, an exact solution may still be acquired by variable separation.[70]

⊃**70** Even if the external system is non-linear, the exact solution can be derived by variable separation method (applicable to 4th item of the 1st problem of this chapter's exercise).

2.3.1   Population dynamics: Logistic curve

Recall the self-reproducing behavior of the optimistic species in the test tube as represented by the ODE in eq. 2.8a and the exact solution in eq. 2.11a. Also, assume no influence from the external system ($b_2 = 0$). Since we have assumed the reproduction rate (a_2) to be unchanging, the system manifests a positive feedback resulting in an exponential increase with time. However, in the real environment, food and living space restrictions should be considered. Overcrowding can occur inside the test tube. With time, reproduction slows down (i.e. reproduction rate decreases). To consider this, eq. 2.8a can be improved to fit a more realistic condition by assuming that the internal parameter a_2 reduces proportionally with the existing microorganisms[71]. a_2 will now be expressed as shown in eq. 2.27

⊃**71** It is important to simplify internal parameter considerations for negative feedback as functions of internal system interactions and not from various external systems. Negative feedback can be set through this internal parameter as a function of the increasing value of dependent variable. This is an excellent feature of this program.

$$a_2 \rightarrow a_3 - a_4 N \qquad (2.27)$$

The influence of the updated internal parameter can be understood by referring to the behavioral effect of the internal parameter in the linear ODE (Table 2.1). If the existing number of individual species N is small, the system will behave similar to the optimistic species. On the contrary when N is large, the system will behave similar to the pessimistic species. When the number increases further in time, the rate of increase of N will attenuate (i.e. $\mathrm{d}N/\mathrm{d}t \rightarrow 0$) and the value of N will approach a_3/a_4 (Table 2.2).

Substituting eq. 2.27 to eq. 2.8a and solving the exact solution through variable separation method,

$$\frac{\mathrm{d}N}{\mathrm{d}t} = (a_3 - a_4 N)N$$

Table 2.2   Influence of the internal parameter $a_3 - a_4N$ to the solution ($b_2 = 0$)

$a_3 - a_4N_0 > 0$ (positive growth rate)	$a_3 - a_4N > 0$	N converges to a_3/a_4 while increasing in time
$a_3 - a_4N_0 = 0$ (no growth rate)	$a_3 - a_4N = 0$	$dN/dt = 0$ or N is unchanging
$a_3 - a_4N_0 < 0$ (negative growth rate)	$a_3 - a_4N < 0$	N converges to a_3/a_4 while attenuating with time

$$\int \frac{1}{(a_3 - a_4N)N}\,dN = \int dt \quad ^{\circ\text{72}}$$

$$\frac{1}{a_3}\int \left(\frac{a_4}{a_3 - a_4N} + \frac{1}{N}\right)dN = \int dt \quad ^{\circ\text{73}}$$

$$\frac{1}{a_3}\left(-\ln|a_3 - a_4N| + \ln|N|\right) + C'' = t$$

$$\frac{N}{a_3 - a_4N} = C'\exp(a_3t)$$

Exact solution: $N = \dfrac{a_3}{a_4 + C\exp(-a_3t)}, \quad C = \dfrac{a_3 - a_4N_0}{N_0}.$

◦**72**   Variable separation

◦**73**   Convert to partial fractions to ease integration process.

From the exact solution, it can be seen that reproduction (i.e. population increase) and destruction (i.e. population decrease) depends on whether the initial population is smaller or greater than a_3/a_4. Finally, as $t \to \infty$, the number of organisms will approach $N = a_3/a_4$. This is called a logistic curve.

2.4   Exercise 1

2.4.1   Investigating propagation of species

Organisms of the same species are living in a controlled environment. They reproduce to increase their population N. The species' propagation in the controlled environment was found to follow the logistic curve. It is an equation of the 1st order ODE (as discussed in the previous lecture) represented by ,

$$\frac{dN}{dt} = (a_3 - a_4N)N.$$

a_3 and a_4 are internal parameters of the system. Based on previous experiments, a_3 (growth rate) is known to be $= 0.05/\text{day}$, and a_4 (carrying capacity) $= 0.00005/\text{organism/day}$.

Your task is to investigate the daily change of N by constructing a time-series plot. You are given the following initial conditions of $N_{t=0} = N_0$ (Dirichlet condition):

1. $N_0 = 100$ organisms

2. $N_0 = 2000$ organisms

2.4.2   Procedures for solving Exercise 1

Unlike the tutorial which focuses on interactive mode, a script may be constructed and stored into a file with an extension '.py'. The script is then executed using the command, 'python (filename).py'. Begin by typing the commands to import the necessary libraries matplotlib.pyplot (http://matplotlib.org/api/pyplot_api.html), and numpy (https://docs.scipy.org/doc/numpydev/user/quickstart.html). The following is the code to import the necessary modules (minimum modules needed for this exercise).

```
1   import matplotlib.pyplot as plt
2   import numpy as np
```

From the given problem of the exercise, parameters that are not changing may be initialized. Furthermore, the naming of the variables is up to the program developer but it is advised that the variable should be easy to understand. In this exercise, the variables and parameters may be assigned as follows.

- Parameters: a3, a4
- Dependent, independent variables: N,t
- Boundary or initial condition: Ninit

The variables are then written into the program script as follows. To construct array variables using NumPy, we can use np.arange (https://docs.scipy.org/doc/numpy/reference/generated/numpy.arange.html) and np.zeros (https://docs.scipy.org/doc/numpy/reference/generated/numpy.zeros.html).

```
1    #parameters
2    a3=0.05
3    a4=0.00005
4
5    #dependent & independent variables
6    t = np.arange(0,120,1)
7    #temporarily construct an array of size similar to
         t.
8    N=np.zeros(t.shape,dtype=np.float32)
9
10   #boundary condition
11   Ninit=100
```

After defining the variables, the exact solution is then encoded to determine the values of N for each t as follows. Observe how the code resembles the written equation in the earlier discussion. Furthermore, notice how the exponent function is simply expressed by "np.exp" where np corresponds to the object which refers to the NumPy library. The exp function from np is used to do exp operation for each individual element of t.

```
1   C=(a3-a4*Ninit)/Ninit
2   N=a3/(a4+C*np.exp(-a3*t))
```

After encoding N, the time-series of N can now be plotted as follows.

```
1   plt.plot(t,N) #t - horizontal. N - vertical
2   plt.show() #Display the results to a new window.
```

The following figure is produced.

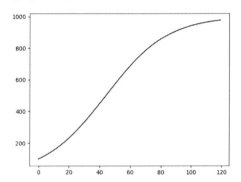

To improve the figure, labels may be added for ease of interpreting the results. The following codes may be appended before the line "plt.show()."

```
1   plt.plot(t,N)
2   #The earlier plot is to be created again
3   #because plt.show() tends to clear the figure
      after showing.
4   plt.xlabel("Days")
5   plt.ylabel("Population")
6   plt.show()
```

The following figure is produced which is similar to the earlier figure but with more detailed information.

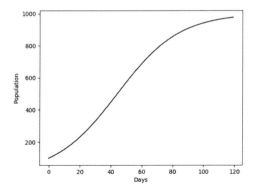

The second case may be overlain to the previous figure by inserting the following code prior to "plt.show()."

```
1   Ninit=2000
2   C=(a3-a4*Ninit)/Ninit
3   N=a3/(a4+C*np.exp(-a3*t))
4   plt.plot(t,N) # Add a plot for the second case
        with Ninit=2000
```

The following figure is produced which is similar to the earlier figure but with the second case.

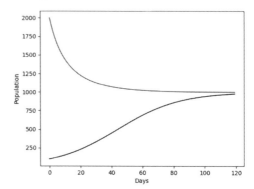

Finally, the figure is improved by inserting legends such as in the following full code for plotting.

```
1   Ninit=100
2   C=(a3-a4*Ninit)/Ninit
3   N=a3/(a4+C*np.exp(-a3*t))
4   plt.plot(t,N,'r-',label='100')
5   plt.xlabel("Days")
6   plt.ylabel("Population")
7   plt.title('Time series')
8   Ninit=2000
9   C=(a3-a4*Ninit)/Ninit
10  N=a3/(a4+C*np.exp(-a3*t))
11  plt.plot(t,N,'b-',label='2000') # Add a plot for
        the second case with Ninit=2000
12  plt.legend(loc=4)
13  plt.show()
```

The following figure is as follows.

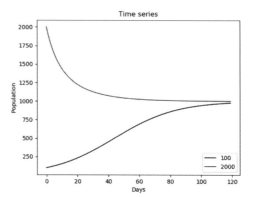

Comparing the time-series plots for two cases with the actual exact solution, the logistic behavior manifested by the species being investigated can be seen more clearly.

◆◆◆◆◆◆ **Problems for Exercise** ◆◆◆◆◆◆

2.1 Find the general solution of the following 1st order ordinary differential equation using the variable separation method. State whether the ODE is linear or non-linear.

1. $\dfrac{\mathrm{d}T}{\mathrm{d}t} = \ln t$ ⤳**74**

2. $\dfrac{\mathrm{d}T}{\mathrm{d}t} = \dfrac{1}{t^2 - 9}$ ⤳**75**

3. $\dfrac{\mathrm{d}T}{\mathrm{d}t} = t^2 T$ ⤳**76**

4. $t^2 \dfrac{\mathrm{d}T}{\mathrm{d}t} = 4 - T^2$ ⤳**77**

5. $\dfrac{\mathrm{d}T}{\mathrm{d}t} = t \cos t$ ⤳**78**

2.2 Find the general solution of the following 1st order ODE using the variable separation method. Although it seems the method will not apply on the first step, it can be applied after variable conversion ⤳**79**.

1. $\dfrac{\mathrm{d}T}{\mathrm{d}t} = \dfrac{T}{t} + 3$

2. $\dfrac{\mathrm{d}T}{\mathrm{d}t} = 2\dfrac{T}{t} + \dfrac{t}{T}$

2.3 A mass of 1 kg is dropped at an initial velocity of 0 m/s. If the air resistance is proportional to the square of the velocity, find the exact solution to predict the time-change in the mass' velocity and graph it. Furthermore, given the resistance coefficient to be 9.8 kg/m and the gravitational acceleration as 9.8 m/s^2, what is the terminal velocity?

2.4 A mass of 1 kg is dropped at an initial velocity of 0 m/s. If the air resistance is proportional to velocity, find the exact solution to predict the time-change in the mass' velocity and graph it. Furthermore, given the resistance coefficient to be 9.8 kg/m and the gravitational acceleration as 9.8 m/s^2, what is the terminal velocity? Compare this result with the previous problem.

2.5 There is a theory that dinosaurs have higher body heat retention than smaller animals because of its large size. Explain this result using the box models introduced in this chapter. Assume generation (source) and loss (sink) due to metabolism, heat release due to sweating are not considered at all, and that body heat release occurs only due to the temperature difference

⤳**74** $\int \mathrm{d}T = \int \ln t \, \mathrm{d}t$
Partially integrate the right side.

⤳**75** $\int \mathrm{d}T = \int 1/(t^2 - 9) \, \mathrm{d}t$
Integrate after partial fraction decomposition of the right-side.

⤳**76** $\int (1/t) \mathrm{d}T = \int t^2 \mathrm{d}t$

⤳**77** $\int \{1/(4 - T^2)\} \mathrm{d}t = \int 1/t^2 \mathrm{d}t$
Integrate after partial fraction decomposition of the left-side.

⤳**78** $\int \mathrm{d}T = \int t \cos t \, \mathrm{d}t$
Partially integrate the right-side.

⤳**79** 1st order ODEs which can be converted from the form $f(aT, at) = f(T, t)$ (a is a real number) to $T' = f(T/t)$ or vice-versa is of homogeneous type. This is because the order of the dependent and independent variables are equal for each term. As will be taught in the next chapter, the second and higher-order ODEs are judged homogeneous based only on the order of the dependent variable (different approach from the previous statement's).

of the body and the external environment.

2.6   The attenuation coefficient of solar radiation in a certain forest is set to a constant value of 0.1 m^{-1}. When solar radiation was measured at bottom ground level of the forest, the solar radiation was found to be 10% of the sky's. Estimate the height of the trees of the forest.

2.7   By examining the existing amount of a radioactive element at two different times, the natural rate of decay of the element can be estimated. Explain the method of estimation.

3 2nd order linear ODE (natural vibration)

Physical phenomena which can be represented by 2nd order linear homogeneous ODE will be discussed. Examples include motion of a mass point connected to a spring with resistance, disk rotation experiencing torsion and resistance, and an RCL circuit. Changing the physical property values (corresponding to the parameters of ODEs) of the parts that constitute the system, whether mechanical or electrical devices, changes the behavior of the system. Unlike 1st order ODEs where the equation representing the solution does not change, the parameters influence multiple possible equations or solutions of the 2nd order ODE. Specifically, the behavior of the system can be classified into 3 types corresponding to 3 possible solutions (real solution, imaginary solution, and multiple roots) of a quadratic equation called "characteristic equation." The focus will be on the application of this concept to the understanding of natural vibrations.

3.1 Method to deriving an exact solution

Given a dependent variable T, the independent variable t, and internal parameters a and b[▷**80**], the 2nd order linear homogeneous ODE can be generalized as follows,

$$\frac{d^2 T}{dt^2} + a\frac{dT}{dt} + bT = 0 \tag{3.1}$$

From the first chapter, the superposition principle can be applicable to eq. 3.1 such that a solution is equal to the sum of its particular solutions as represented by the following equation[▷**81**],

$$T = C_1 T_1 + C_2 T_2 \tag{3.2}$$

▷**80** In general, the parameters of the 2nd order linear homogeneous ODE are not necessarily constant values, but may be functions of independent variables. In this chapter, they are assumed constant.

▷**81** In ch. 1, the superposition principle was briefly introduced. A general form such as eq. 3.2 which depicts an arbitrary sum of two solutions is referred to as "linear combination." Substituting eq. 3.2 into eq. 3.1,
$d^2(C_1 T_1 + C_2 T_2)/dt^2 + a\,d(C_1 T_1 + C_2 T_2)/dt$
$+b(C_1 T_1 + C_2 T_2)$
$= C_1 [d^2 T_1/dt^2 + a\,dT_1/dt + bT_1]$
$+C_2 [d^2 T_2/dt^2 + a\,dT_2/dt + bT_2]$
$= 0$

> Proof of applicability of the superposition principle:
>
> $$\frac{d^2(C_1T_1+C_2T_2)}{dt^2}+a\frac{d(C_1T_1+C_2T_2)}{dt}+b(C_1T_1+C_2T_2)=0$$
>
> $$C_1\left(\cancelto{0}{\frac{d^2T_1}{dt^2}+a\frac{dT_1}{dt}+bT_1}\right)+C_2\left(\cancelto{0}{\frac{d^2T_2}{dt^2}+a\frac{dT_2}{dt}+bT_2}\right)=0$$
>
> Here, C_1 and C_2 are unspecified constants (real or imaginary) that can be determined from the initial and boundary conditions.

▷**82** There is no guarantee that the solution will have exponential terms. This intuition is based on the exponential terms of the solutions of the 1st-order ODEs of the previous chapter and the convenience of differentiating exponential functions.

Let us assume an exponential function as a solution of the ODE▷**82**. The solution is similar in form as the solution of the 1st order ODE (see ch. 2).

$$T = \exp(\lambda t) \quad (= e^{\lambda t}) \tag{3.3}$$

The 1st and 2nd derivative of eq. 3.3 are

$$T' = \lambda \exp(\lambda t) \tag{3.4}$$

$$T'' = \lambda^2 \exp(\lambda t) \tag{3.5}$$

Substituting eq. 3.3 to eq. 3.5 into eq. 3.1 results in,

$$(\lambda^2 + a\lambda + b)\exp(\lambda t) = 0 \tag{3.6}$$

By logical thinking, it can be seen that λ can be solved using

$$\lambda^2 + a\lambda + b = 0 \tag{3.7}$$

▷**83** Ensure that when $e^{(p+iq)t}$, $e^{(p-iq)t}$ are bases of the solution, $e^{pt}\cos qt$, $e^{pt}\sin qt$ are also bases of the solution.
According to the Euler's formula,
$$e^{iat} = \cos at + i\sin at$$
$$e^{-iat} = \cos at - i\sin at$$
Applying the equation above to $e^{(p+iq)t}, e^{(p-iq)t}$,
$$T_a = e^{(p-iq)t} = e^{pt}(\cos qt - i\sin qt)$$
$$T_b = e^{(p+iq)t} = e^{pt}(\cos qt + i\sin qt)$$
The linear combination of T_a and T_b should also be a solution.
$$T_1 = (T_a + T_b)/2 = e^{pt}\cos qt$$
$$T_2 = (T_a + T_b)/2i = e^{pt}\sin qt$$

This quadratic equation is referred to as the **characteristic (auxiliary) equation** of the ODE (eq. 3.1).

The solution of the characteristic equation can be easily acquired using the quadratic formula,

$$\lambda = \frac{-a \pm \sqrt{a^2 - 4b}}{2} \tag{3.8}$$

From algebra, eq. 3.8 can have various solutions depending on the condition of λ: a real double root $\lambda = \alpha$, distinct real roots $\lambda = \alpha, \beta$, and a imaginary root of the form $\lambda = p \pm iq$. The value of λ will serve as basis for the solution of ODE. It is not easy to physically imagine an imaginary number so it is better to convert it into a trigonometric function. The summary is shown in Table 3.1▷**83**.

Table 3.1 The general equation of the ODE (eq. 3.1) based on the classes of solutions to the characteristic equation

	distinct real roots $(\lambda = \alpha, \beta)$	real double root $(\lambda = \alpha)$	imaginary root $(\lambda = p \pm iq)$
Basis	$e^{\alpha t}, e^{\beta t}$	$e^{\alpha t}, te^{\alpha t}$	$e^{pt}\cos qt, e^{pt}\sin qt$
General Solution	$T = C_1 e^{\alpha t} + C_2 e^{\beta t}$	$T = C_1 e^{\alpha t} + C_2 t e^{\alpha t}$	$T = e^{pt}(C_1\cos qt + C_2\sin qt)$

Let us examine the detailed derivation for each general solution. The aim is to derive a general solution which is equivalent to the sum of two particular solutions (or basis of solutions) and the quotient of both is not constant (i.e. one particular solution must not be divisible by another particular solution).

1. Distinct real roots

If $a^2 > 4b$ (note that a and b are internal parameters) in eq. 3.8, λ can either be equal to

$$\alpha = \frac{-a + \sqrt{a^2 - 4b}}{2} \quad \text{and} \quad \beta = \frac{-a - \sqrt{a^2 - 4b}}{2}$$

Substituting to eq. 3.3 will provide particular solutions (henceforth, we collectively call them as "basis of solutions"),

$$T_1 = \exp(\alpha t); T_2 = \exp(\beta t)$$

The basis of solutions is defined (and real) for all t and the quotient of each particular solution is not constant. Applying the superposition principle (eq. 3.2), the general solution of the ODE (3.1) can be equal to

$$\boxed{T = C_1 \exp(\alpha t) + C_2 \exp(\beta t)}$$

2. Real double root

If $a^2 = 4b$ (note that a and b are internal parameters) in eq. 3.8, λ can be equal to $\alpha = -a/2$. Only one solution can be derived,

$$T_1 = \exp(\alpha t) = \exp\left(-\frac{a}{2}\right) \tag{3.9}$$

Another second independent solution T_2, another approach is needed. Method of "reduction of order" is introduced. The purpose of this method is to derive a second linearly independent solution T_2 by solving a 1st order ODE (one order less than the 2nd order thus reduction of order or both are linearly independent of each other). The procedure begins by setting $T_2 = zT_1$. Differentiating T_2 results in

$$T_2' = zT_1' + z'T_1$$
$$T_2'' = zT_1'' + T_1'z' + z'T_1' + T_1z''$$
$$= zT_1'' + 2z'T_1' + z''T_1$$

Since T_2 is an independent solution, substituting T_2 and its derivatives into eq. 3.1 (can be written as $T'' + aT' + bT = 0$) results in

$$(zT_1'' + 2z'T_1' + z''T_1) + a(zT_1' + z'T_1) + b(zT_1) = 0$$

Grouping terms by z coefficients,

$$z''(T_1) + z'(2T_1' + aT_1) + z(\overbrace{T_1'' + aT_1' + bT_1}^{0}) = 0 \tag{3.10}$$

Substituting eq. 3.9 and $a = -2\alpha$ into eq. 3.10,

$$z''(e^{\alpha t}) + z'(2\alpha e^{\alpha t} + -2\alpha e^{\alpha t}) = 0$$

Simplifying the above equation by dividing it by $e^{\alpha t}$ results in

$$z'' + z'(\underbrace{2\alpha + -2\alpha}_{0}) = 0$$

Notice that the equation is first order in $x = z'$ (thus the name reduction in order),

$$x' = 0 \qquad\qquad (3.11)$$

x of eq. 3.11 can be solved using variable separation (previous chapter) (note that there is no need for a constant of integration because particular solution is aimed) or simply assign 1 or any value,

$$x = 1$$

Solving z by substituting back x,

$$z' = 1$$
$$z = t$$

Finally, another particular solution T_2 can be obtained by substituting z into $T_2 = zT_1$,

$$T_2 = t\exp(\alpha t) \qquad\qquad (3.12)$$

Notice that dividing T_2 by T_1 will result in t (not a constant) making both solutions linearly independent, thus allowing us to derive a general solution (in the form eq. 3.2),

$$\boxed{T = C_1\exp(\alpha t) + C_2 t\exp(\alpha t)}$$

3. Imaginary roots
If $a^2 < 4b$ (note that a and b are internal parameters) in eq. 3.8, λ can be equal to

$$\lambda = \frac{-a \pm \sqrt{-1}\sqrt{4b - a^2}}{2} = -\frac{a}{2} \pm i\frac{\sqrt{4b - a^2}}{2}$$

$$= p \pm iq \qquad (\text{where: } \quad p = -\frac{a}{2} \quad \text{and} \quad q = \frac{\sqrt{4b - a^2}}{2})$$

T_1 and T_2 can *initially* be set to,

$$T_1 = e^{(p+iq)t} \quad \text{and} \quad T_2 = e^{(p-iq)t}$$

Applying the superposition princile such as in eq. 3.2,

$$T = Ae^{(p+iq)t} + Be^{(p-iq)t}$$

where A and B are the arbitrary constants (real and imaginary).
We can simplify the general solution further by separating the

imaginary number i from the particular solutions and including it within the constant coefficients.

$$T = Ae^{pt}e^{iqt} + Be^{pt}e^{-iqt}$$

Recall Euler's formula:

$$e^{ix} = \cos x + i\sin x; \quad e^{-ix} = \cos x - i\sin x$$

Applying Euler's formula and separating e^{pt},

$$T = e^{pt}(A\cos qt + Ai\sin qt + B\cos qt - Bi\sin qt)$$
$$= e^{pt}[(A+B)\cos qt + (A-B)i\sin qt]$$

Note that by definition of superposition principle, A and B can be any arbitrary constants. We can then assume that $A = B$, this will result in a new particular solution for T, say T_3,

$$T_3 = e^{pt}2A\cos qt$$

Correspondingly, if we set $B = -A$, another particular solution can be acquired, say T_4,

$$T_4 = e^{pt}2Ai\sin qt$$

We can once again invoke the superposition principle such that,

$$T = C_6 T_3 + C_7 T_4$$
$$= C_6 2Ae^{pt}\cos qt + C_7 2Aie^{pt}\sin qt$$

The coefficients $C_6 2A$ and $C_7 2Ai$ can be grouped separately by arbitrary constants C_1 and C_2,

$$C_1 = C_6 2A \quad \text{and} \quad C_2 = C_7 2Ai$$

This leads to the general solution,

$$\boxed{T = C_1 \exp(pt)\cos qt + C_2 \exp(pt)\sin qt}$$

3.2   Mass–spring system

Figure 3.1 shows an example of a mass–spring system of two configurations: one with a damper and another without a damper. The mass is attached at one end to an ordinary coil spring that resists expansion and compression. The other end is attached to a damper with a purpose to reduce vibration (e.g. dampers of buildings reduce the vibrations brought about by the shaking of the ground). The quality of the spring and the dampers are represented by internal parameters, k and c, respectively. The

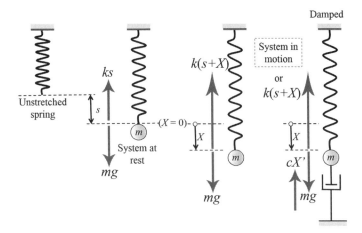

Figure 3.1   Mechanical mass–spring system with internal forces acting on
the mass

reference location is at the free-end of the spring when unattached to the
mass. The displacement s of the string is measured from the state when the
spring and the mass balances out or when the mass is hanging motionless
from the string. A relationship can be established between the weight of the
mass mg and the restoring force of the spring ks,

$$mg = ks. \tag{3.13}$$

The mass is shifted vertically by pulling the mass downwards up to a
distance $X_t = 0$ and then released. At the moment of release, restoring
force from the spring (proportional to the displacement), resistance from
the dampers (proportional to the velocity of the mass), and the gravitational
force will act on the mass and influence its behavior. From Newton's law,
the sum of forces acting on the mass point will be equal to the net force
equivalent to the mass multiplied by its acceleration (inertia). An equation
representing the mass–spring system is derived as follows,

$$m\frac{\mathrm{d}^2X}{\mathrm{d}t^2} = mg - k(s+X) - c\frac{\mathrm{d}X}{\mathrm{d}t}\,{}^{\circlearrowright}\mathbf{84}. \tag{3.14}$$

⟳**84**   Since the displacement X is
positive in the downward direction,
the sign of gravitational force acting downward (1st term on the right
side) is positive, while the signs
of the spring's restoring force (2nd
term) and that of the resistance
force (3rd term) are negative.

Substituting eq. 3.13 into eq. 3.14 will result in a 2nd order linear homogeneous ODE with X as the dependent variable and t as the independent
variable,

$$m\frac{\mathrm{d}^2X}{\mathrm{d}t^2} + c\frac{\mathrm{d}X}{\mathrm{d}t} + kX = 0\,{}^{\circlearrowright}\mathbf{85} \tag{3.15}$$

⟳**85**   Each of the three constants
will have the following units for
consistency:
$m[\mathrm{kg}], c[\mathrm{kg/s}], k[\mathrm{kg/s}^2]$

It is interesting that when comparing eq. 3.15 above with eqs. 2.5, 2.7,
gravity as an external force is no longer considered. The system consists of
only three internal parts namely the inertial force, the restoring force, and
the resistance force. This means that even if the system is brought to the
moon from the earth where gravity is largely different (of course, excluding
the influence of air resistance of the mass and the spring), the behavior will
still be the same.

Although the internal parameters of eq. 3.15 appear to be three (mass m, spring constant k, and resistance coefficient c), dividing the terms by m will reduce the parameters into two.

$$\frac{d^2X}{dt^2} + c_p \frac{dX}{dt} + \omega_p^2 X = 0 \quad \text{where} \quad c_p = c/m, \quad \omega_p = \sqrt{k/m} \quad (3.16)$$

The reason for the definition of the second internal parameter ω_p will be seen in the further discussion. One obvious reason will be for both c_p and ω_p to have the same physical units of $1/s$.

Even if the combination of the three physical properties (m, c, k) are different for the two systems, as long as the internal parameters c_p and ω_p are the same, both systems will be identical.

Let us find the exact solution of the ODE (eq. 3.16). Assuming an initial displacement and velocity to be $X_{t=0} = X_0$, and $X'_{t=0} = 0'$ (Neumann condition), the characteristic equation (as discussed earlier) can be expressed as follows,

$$\lambda = \frac{-c_p \pm \sqrt{c_p^2 - 4\omega_p^2}}{2}. \quad (3.17)$$

Similar to Sect. 3.1, depending on the set-up of the internal parameters, various conditions may be investigated based on the derived ODE (eq. 3.16). Here we will investigate undamped and damped systems. Damped systems can be sub-divided further into: underdamping, critically damped, and overdamped. To derive the solution, Table 3.1 will be used.

1. Undamped system $(c_p = 0)$

First, let us consider the system when the dampers are not attached or not functioning. Note that in reality, every system has damping. Otherwise, a system will keep moving forever. Here, damping is assumed to be so small and the analysis is conducted only for a relatively short time.

$$\lambda = \pm i\omega_p \quad (3.18)$$

The characteristic equation (eq. 3.18) is an imaginary solution. From Table 3.1, $p = 0$ and $q = \omega_p$. A general solution can be derived as,

$$X = C_1 \cos \omega_p t + C_2 \sin \omega_p t \quad (3.19)$$

From the initial conditions above, $C_1 = X_0$ and $C_2 = 0$[86]. The exact solution is,

$$X = X_0 \cos \omega_p t. \quad (3.20)$$

The resulting motion is called a **harmonic oscillation** with an amplitude of X_0 and a **natural angular frequency** of $\omega_p = \sqrt{k/m}$[87]. The periodic motion is the **natural oscillation** of the system. What makes the system "unique" is without the action of the external system, periodic motion occurs simply through the balance of inertial force and the restoring force of the spring, and its angular frequency is just determined by the internal parameters.

[86] Since $X' = \omega_p(-C_1 \sin \omega_p t + C_2 \cos \omega_p t)$, $X_0 = C_1$, $X'_{t=0} = 0 = \omega_p C_2$

[87] Since the period of eq. 3.23 is $2\pi/\omega_p[s]$, the frequency is its reciprocal ($\omega_p/2\pi[1/s]$) which describes the number of cycles per unit time.

2. Damped system: Underdamping ($c_p < 2\omega_p$)

The solution of the characteristic equation (eq. 3.17) becomes an imaginary solution when the internal parameter c_p (damper) becomes smaller than ω_p (spring). When the damping resistance was 0, the mass naturally oscillates forever. For the current case where slight resistance is applied, what is the behavior of the mass?

$$\lambda = p \pm iq, \qquad p = \frac{-c_p}{2}, \qquad q = \frac{\sqrt{4\omega_p^2 - c_p^2}}{2} \qquad (3.21)$$

The characteristic equation (eq. 3.21) suggests an imaginary solution. According to Table 3.1, the general solution will be as follows,

$$X = e^{pt}(C_1 \cos qt + C_2 \sin qt). \qquad (3.22)$$

▷**88** $X' = pe^{pt}(C_1 \cos qt + C_2 \sin qt)$
$+ e^{pt} q(-C_1 \sin qt + C_2 \cos qt)$,
$X_0 = C_1$,
$X'_{t=0} = pC_1 + qC_2 = 0$

From the initial conditions, $C_1 = X_0$ and $C_2 = -X_0 p/q$▷**88** can be derived. Substituting the constants to eq. 3.22, the exact solution is,

$$X = X_0 e^{pt}\left(\cos qt - \left(\frac{p}{q}\right)\sin qt\right) \qquad (3.23)$$

▷**89** The internal parameters are all physical coefficients that are always positive.

▷**90** $p = -C_p/2$

▷**91** $q = \sqrt{4\omega_p^2 - c_p^2}/2$
$= \sqrt{\omega_p^2 - (c_p/2)^2}$
$< \sqrt{\omega_p^2} = \omega_p$

Analyzing eq. 3.23 together with the fact that c_p, ω_p are positive values▷**89**, it can be seen that p will always be a negative value▷**90**. Furthermore the value of q must always be smaller than the natural frequency ω_p▷**91**. For the sake of simplicity, we assume that the damping resistance is much smaller than the spring coefficient such that the right-hand term $((p/q)\sin qt)$ of eq. 3.23 becomes negligible. For this system, the right-hand side of eq. 3.23 becomes the product $X_0 e^{pt}\cos qt$. This results in a vibration in the shape of a cosine and exponentially attenuating with time (since p is negative). The mass motion should gradually attenuate to the static equilibrium state (motionless) by reducing the amplitude while oscillating. Therefore, the mass point corresponding to the solution eq. 3.23 is generally called a **damped oscillation**. Since q is smaller than the natural angular frequency ω_p with increasing damping resistance, the oscillation is slower and the attenuating speed (as defined by p) becomes faster (i.e. reaches static equilibrium earlier).

3. Damped system: Critical damping ($c_p = 2\omega_p$)

Let us assume that the internal parameter of the damping resistance is equivalent to twice the natural angular frequency. The characteristic equation (eq. 3.17) becomes,

$$\lambda = \alpha = \frac{-c_p}{2} = -\omega_p \qquad (3.24)$$

The general solution of eq. 3.17) will be based upon a real double root (Table 3.1),

$$X = C_1 e^{\alpha t} + C_2 t e^{\alpha t} \qquad (3.25)$$

▷**92** $X' = \alpha C_1 e^{\alpha t} + \alpha t C_2 e^{\alpha t} + C_2 e^{\alpha t}$,
$X_0 = C_1$,
$X'_0 = \alpha C_1 + C_2 = 0$,

From the initial conditions, $C_1 = X_0$ and $C_2 = -\alpha X_0$▷**92** can be derived. Substituting the constants to eq. 3.25 will result in,

$$X = X_0(1 - \alpha t)e^{\alpha t} = X_0(1 + \omega_p t)e^{-\omega_p t} \qquad (3.26)$$

This is the solution of the critical condition when the resistance value is gradually increased in the solution of the damped oscillation and decreased in the solution of the overdamped oscillation to achieve the fastest attenuation. This condition is called **critical damping**. If a system is desired which can attenuate the displacement X (or vibration) at the earliest time, a system can be constructed by adjusting the internal parameters to satisfy critical damping (also see Table 3.2).

4. Damped system: Overdamping $(c_p > 2\omega_p)$

The roots of the characteristic equation (eq. 3.17) becomes two real distinct roots,

$$\lambda = \alpha, \beta, \qquad \alpha = \frac{-c_p + \sqrt{c_p^2 - 4\omega_p^2}}{2}, \qquad \beta = \frac{-c_p - \sqrt{c_p^2 - 4\omega_p^2}}{2} \tag{3.27}$$

Since the characteristic equation (eq. 3.27) satisfies the condition of double real roots (Table 3.1), the general solution will be

$$X = C_1 e^{\alpha t} + C_2 e^{\beta t} \tag{3.28}$$

From the initial conditions, $C_1 = -\beta X_0/(\alpha - \beta)$ and $C_2 = \alpha X_0/(\alpha - \beta)$[93] can be derived. Substituting the constants to eq. 3.25 will result in,

$$X = \frac{X_0}{\alpha - \beta}(-\beta e^{\alpha t} + \alpha e^{\beta t}) \tag{3.29}$$

Considering the behavior of this solution, it must be recognized that the internal parameters c_p, ω_p are always positive[94], so that α and β of eq. 3.27 should be correspondingly negative[95]. In other words, the exact solution is a superposition of two exponentially decaying curves $-\beta e^{\alpha t}, \alpha e^{\beta t}$ of varying attenuation speed and amplitude. In particular, the former $(-\beta e^{\alpha t})$ will have a more dominant effect to the solution simply because the absolute value of its amplitude is larger $(|\beta| > |\alpha|)$, and the rate of time of decay is slower $(e^{\alpha t} \geqq e^{\beta t})$. This results in the mass point gradually approaching the zero displacement $(X = 0)$ with lesser or no oscillation since its initial displacement X_0. This condition is called **overdamping**. With larger damping resistance, attenuation is actually slower[96]. In other words, the damping effectively hinders movement of the mass back to the zero displacement (i.e. takes more time to reach static equilibrium).

A summary of the exact solution is shown in Table 3.2.

▷**93** $X' = \alpha C_1 e^{\alpha t} + \beta C_2 e^{\beta t}$,
$X_0 = C_1 + C_2$,
$X'_{t=0} = \alpha C_1 + \beta C_2 = 0$

▷**94** The internal parameters represent physical quantities which are always positive.

▷**95** It is obvious that $\beta < 0$.
$\alpha = -(c_p/2)$
$\quad + \sqrt{(c_p/2)^2 - \omega_p^2}$
$\quad\quad < -(c_p/2) + \sqrt{(c_p/2)^2} = 0$

▷**96** If the value of the resistance coefficient c_p increases, the α of negative value becomes closer to 0 resulting in slower time for attenuation.

Table 3.2 Summary of solutions for eq. 3.16 and corresponding physical phenomena[97]

Solution type	Form of Solution	Physical Phenomena
Imaginary $(c_p = 0)$	$X = X_0 \cos \omega_p t$ $(\lambda = \pm i\omega_p)$	Natural vibration
Imaginary $(c_p < 2\omega_p)$	$X = X_0 e^{pt}(\cos qt - (p/q)\sin qt)$ $(\lambda = p \pm iq)$	Damped: Underdamping
Double root $(c_p = 2\omega_p)$	$X = X_0(1 - \alpha t)e^{\alpha t}$ $(\lambda = \alpha)$	Damped: Critical damping
Distinct roots $(c_p > 2\omega_p)$	$X = X_0(\alpha - \beta)^{-1}(-\beta e^{\alpha t} + \alpha e^{\beta t})$ $(\lambda = \alpha, \beta)$	Damped: Overdamping

▷**97** The difference in functional forms of the solution of the ODE corresponds to varying physical phenomenon. Even if systems describe the same physical phenomena, the physical property values determines individuality (i.e. difference in internal parameters, may lead to a different phenomenon or mathematical solution).

3.3   Behavior of current flowing in an RCL circuit

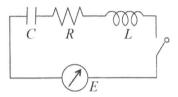

Figure 3.2   RCL electric circuit

Consider an electric circuit where a coil (i.e. resistor) of resistance R [unit:$\Omega = V/A$], capacitor of capacitance C [unit:As/V], and inductor of inductance L [unit:Vs/A] are connected in series with a power supply of voltage E [unit:V]. This is commonly called an **RCL circuit** (Figure 3.2). The oscillation of the current in the RCL electric circuit can be represented by a 2nd order ODE as the above natural oscillation of the spring. When the circuit is switched on, current I will begin to flow in the circuit. The voltage drop brought about by the inductor is $V_L = LI'$. Following Kirchhoff's voltage law (KVL)[99][98], the following formula can be established,

> [99][98]   Kirchhoff's voltage law states, "In any closed loop network, the total voltage around the loop is equal to the sum of all the voltage drops within the same loop."

$$L\frac{\mathrm{d}I}{\mathrm{d}t} + RI + \frac{Q}{C} = E \tag{3.30}$$

Electric charge Q can be related with the current using eq. 3.32. Substituting eq. 3.32 to 3.30 and differentiating with respect to time t will result in a 2nd order ODE (eq. 3.33).

$$I = \frac{\mathrm{d}Q}{\mathrm{d}t}, \tag{3.31}$$

> [99]   Similar to the mass–spring system, the constant external electromotive force vanishes from the ODE. The phenomenon is only affected by the internal system.

$$\frac{\mathrm{d}^2 I}{\mathrm{d}t^2} + \left(\frac{R}{L}\right)\frac{\mathrm{d}I}{\mathrm{d}t} + \left(\frac{1}{L}\frac{1}{C}\right)I = 0,^{[99]} \tag{3.32}$$

$$\frac{\mathrm{d}^2 I}{\mathrm{d}t^2} + c_p\frac{\mathrm{d}I}{\mathrm{d}t} + \omega_p^2 I = 0. \tag{3.33}$$

From eq. 3.32, it is interesting to see that the contribution of the constant voltage power supply E to the current I vanishes. Instead, the internal system parts, capacitor, resistor, and inductor describes the behavior of the dependent variable (i.e. current). In addition, the internal parameters R, L, C are less important than the ratios as shown in eq. 3.32 where the internal parameters can be regrouped into two; the $c_p = (R/L)$[100] and $\omega_p = \sqrt{1/LC}$. Both c_p and ω_p have units $1/s$. ω_p can also be referred to as the natural angular frequency.

> [100]   Units of c_p and ω_p (natural angular frequency) are [1/s] and [1/s], respectively.

Eq. 3.33 and eq. 3.16 are exactly the same type of ODE. Therefore, RCL electric circuits and the mass spring may differ only by appearance, but its behavior can be interpreted in a similar manner.

3.4   Disk rotation subjected to tortion and resistance

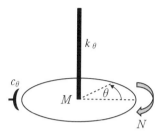

Figure 3.3   Rotating disk hanging from a sturdy string

Consider the case of a disk fixed along a vertical axis (i.e. sturdy string) as shown in Figure 3.3 and experiencing friction at its edges. The disk is twisted by a constant external torque N [unit: $kg \cdot m^2/s^2$][101]. Let θ be the angular displacement or the angle of twisting. Torque N generated by the effort of the vertical axis to return to its original position upon twisting is proportional to the angle θ and acting in the opposite direction. The torsional coefficient (restoring coefficient of the vertical axis) is represented by k_θ. Also, the frictional resistance proportional to the angular velocity generates torque. The factor of proportionality (resistance coefficient) is represented by c_θ [unit: $kg \cdot m^2/s$]. For the linear motion of a given mass, Newton's law (represented by eq. 3.34) is applicable which simply means that the product (inertial force) of the mass m and the acceleration X'' is equal to the sum of the forces acting along the direction of motion. Likewise in the case of rotational motion, the product (inertial force) of the moment of inertia M [unit: $kg \cdot m^2$] and the angular acceleration θ'' is equal to the sum of the torques N acting along the direction of rotation (eq. 3.35). Here, torque is the cross product between the distance r from the center of the axis of rotation to the point of the action and the force F itself.

$\triangleright$**101**   [$N \cdot m$] can be used as the unit of torque. N stands for Newton which is equivalent to $kg \cdot m/s^2$.

$$mX'' = F, \tag{3.34}$$

$$M\theta'' = N = r \times F. \tag{3.35}$$

Let us apply eq. 3.34 to the rotating disk. Considering that both the torque generated by frictional resistance and restoring action of the vertical axis act in opposite direction, the following expression can be obtained. However, the torque (external torque) does not act as an external system,

$$M\frac{d^2\theta}{dt^2} = -c_\theta \frac{d\theta}{dt} - k_\theta\theta, \tag{3.36}$$

$$\frac{d^2\theta}{dt^2} + \frac{c_\theta}{M}\frac{d\theta}{dt} + \frac{k_\theta}{M}\theta = 0, \tag{3.37}$$

$$\frac{d^2\theta}{dt^2} + c_p\frac{d\theta}{dt} + \omega_p^2\theta = 0. \tag{3.38}$$

At first glance, the behavior of the disk seems to be influenced by 3 internal parameters, the moment of intertia M, the resistance coefficient c_θ, and the torsional coefficient k_θ. However, the internal parameters can be reduced further into $c_p = (c_\theta/M)$[102] and $\omega_p = \sqrt{k_\theta/M}$.

▷**102** Units of c_p and ω_p (natural angular frequency) are [1/s] and [1/s], respectively.

Eq. 3.38 reduces to the same type of ODE as the mass–spring system (eq. 3.16). Therefore, it can be interpreted that the rotational motion of the disk experiencing torsion and resistance is a physical phenomenon equivalent to a mass–spring system (mechanical device) and an RCL electric circuit.

An analogy can be confirmed for the 3 systems discussed so far (summarized in Table 3.3). Except for the dependent variables and the description of the system, the normalized 2nd order homogeneous ODE representing the 3 systems will result in the same behavior and defined only by the resistance parameter c_p and the natural angular frequency ω.

Table 3.3   Analogy of mass–spring systems, RCL circuits, and rotary motions[103]

Mass–spring system	RCL circuit	Rotating disk
Displacement X	Current I	Angular displacement θ
Mass m	Inductance L	Moment of inertia M
Frictional resistance c	Electric resistance R	Frictional resistance c_θ
Spring constant k	Reciprocal of capacitance $1/C$	Torsional constant k_θ
Gravitation mg	Electromotive force E	External torque N_{out}
Resistance parameter $c_p = \frac{c}{m}$	Resistance parameter $c_p = \frac{R}{L}$	Resistance parameter $c_p = \frac{c_\theta}{M}$
Natural frequency $\omega_p = \sqrt{\frac{k}{m}}$	Natural frequency $\omega_p = \sqrt{\frac{1}{CL}}$	Natural frequency $\omega_p = \sqrt{\frac{k_\theta}{M}}$
$\frac{d^2X}{dt^2} + c_p\frac{dX}{dt} + \omega_p^2 X = 0$	$\frac{d^2I}{dt^2} + c_p\frac{dI}{dt} + \omega_p^2 I = 0$	$\frac{d^2\theta}{dt^2} + c_p\frac{d\theta}{dt} + \omega_p^2 \theta = 0$

▷**103** When internal parameters are normalized, the resistance coefficient c_p and the natural angular frequency ω_p are both in [1/s] units (dimension of frequency).

3.5   Deriving the law of conservation of energy using the law of conservation of momentum

3.5.1   Derive the law of energy conservation from law of momentum conservation

Up to this point, the law of conservation of momentum has been taken as the physical law which describes the behavior of the mass–spring system. Since the law itself includes acceleration which is a second order derivative of the displacement, there is no difficulty in expressing the system by ODE. Aside from the law of conservation of momentum, the law of conservation of energy[104] is also widely known. Towards the end of this chapter, it will be shown that the law of conservation of energy is derived by integrating the law of conservation of momentum and; conversely, the law of conservation of momentum is derived by differentiating the law of conservation of energy[105].

▷**104** It will be useful to know how the law of energy conversation relates with ODE.

▷**105** The only difference is whether the universal phenomenon of mass motion is expressed in differential form (law of conservation of momentum) or integral form (law of conservation of energy).

Here, we will examine the mass–spring system. In general, the product of force F and displacement dX is work dW.

$$dW = F\,dX$$

When both sides are integrated from the initial displacement $X_{t=0} = X_0$ to the displacement $X_{t=t_1} = X_1$ at a certain time, the amount of work exerted by the system until time t_1 should be obtained. Integrating eq. 3.15 by dX from displacement X_0 to X_1 will result in

$$\int_{X_0}^{X_1} m\frac{d^2 X}{dt^2}\,dX + \int_{X_0}^{X_1} c\frac{dX}{dt}\,dX + \int_{X_0}^{X_1} kX\,dX = 0. \tag{3.39}$$

From the relationship shown in eq. 3.40, eq. 3.39 can be converted to an integration with respect to time. This process is crucial since t is the independent variable.

$$dX = \frac{dX}{dt}\,dt \tag{3.40}$$

Substituting eq. 3.40 into eq. 3.39,

$$\int_{t_0}^{t_1} m\left(\frac{d^2 X}{dt^2}\right)\left(\frac{dX}{dt}\right)dt + \int_{t_0}^{t_1} c\left(\frac{dX}{dt}\right)^2 dt + \int_{t_0}^{t_1} kX\left(\frac{dX}{dt}\right)dt = 0. \tag{3.41}$$

$$\int_{t_0}^{t_1} \frac{1}{2}m\frac{d}{dt}\left(\frac{dX}{dt}\right)^2 dt + \int_{t_0}^{t_1} c\left(\frac{dX}{dt}\right)^2 dt + \int_{t_0}^{t_1} \frac{1}{2}k\frac{d}{dt}\left(X^2\right)dt = 0. \text{▷106} \tag{3.42}$$

> **▷106** Since $df(t)^2/dt$
> $= 2f(t)\,df(t)/dt$,
> $d(dX/dt)^2/dt$
> $= 2(dX/dt)(d^2X/dt^2)$,
> $dX^2/dt = 2X(dX/dt)$.

Setting $v = \frac{dX}{dt}$, $v_0 = \left(\frac{dX}{dt}\right)_{t=t_0}$, $v_1 = \left(\frac{dX}{dt}\right)_{t=t_1}$ results in

$$\frac{1}{2}mv_1^2 - \frac{1}{2}mv_0^2 + \int_{t_0}^{t_1} cv^2\,dt + \frac{1}{2}kX_1^2 - \frac{1}{2}kX_0^2 = 0.$$

Since only the energy change due to the resistance (third term) can not be integrated, let us transpose this term to the right-side of the equation.

$$\underbrace{\frac{1}{2}mv_1^2 - \frac{1}{2}mv_0^2}_{(a)} + \underbrace{\frac{1}{2}kX_1^2 - \frac{1}{2}kX_0^2}_{(b)} = \underbrace{-\int_{t_0}^{t_1} cv^2\,dt}_{(c)} \tag{3.43}$$

> **▷107** Term (c) depends on the function form of the velocity v as function of time, so it cannot be integrated. However, after the solution of the ODE is obtained, an exact solution of v can also be obtained. Substituting the exact solution will allow us to estimate the term (c).

(a) Change in kinetic energy of the mass.

(b) Change in elastic potential energy of the spring.

(c) Energy attenuation by resistance▷107.

If the system exerts no work, kinetic energy and potential energy (related to the spring's elasticity) balances. This equation does not include the potential energy from the mass point. Does the above really depict the law of conservation of energy? Let's see below.

3.5.2 Derive the equation of motion from the law of energy conservation

This time, let us derive the conservation of momentum by differentiating the law of conservation of energy. Since energy attenuation due to resistance cannot be formulated, we shall start from the energy conservation law without resistance. The energy of the system can be broken down into three, the mass's kinetic energy, the potential energy, and the spring's elastic potential energy. Keeping in mind that X is defined to be a distance downward from the equilibrium position[108],

⊳**108**   The sign of potenial energy becomes negative.

$$\frac{1}{2}mv_1^2 + \frac{1}{2}k(X_1 + s)^2 - mgX_1 = \frac{1}{2}mv_1^2 + \frac{1}{2}k(X_1^2 + 2X_1 s + s^2) - mgX_1$$

$$= \frac{1}{2}mv_1^2 + (ks - mg)X_1 + \frac{1}{2}kX_1^2 + \frac{1}{2}ks^2$$

$$= \frac{1}{2}mv_1^2 + \frac{1}{2}kX_1^2 + \frac{1}{2}ks^2 \qquad (3.44)$$

Here, we used the relationship of equilibrium derived earlier, $mg = ks$. The same equation holds for the energy of initial state; so if we take the difference of the sum of the work, the third term of eq. 3.44 cancels and the equation will coincide with the law of energy conservation (eq. 3.43).

$$\frac{1}{2}mv_1^2 + \frac{1}{2}kX_1^2 = \frac{1}{2}mv_0^2 + \frac{1}{2}kX_0^2 = \text{constant} \qquad (3.45)$$

Differentiating eq. 3.45 with respect to independent variable t,

$$\frac{1}{2}2mv\frac{dv}{dt} + \frac{1}{2}2kX\frac{dX}{dt} = mv\frac{dv}{dt} + kXv = v\left[m\frac{dv}{dt} + kX\right] = 0,$$

$$m\frac{d^2X}{dt^2} + kX = 0. \qquad (3.46)$$

From the above, the law of conservation of momentum is expressed without the action of resistance.

3.6   Exercise 2

3.6.1   Investigating the behavior of a mass–spring system without external influence

Four (4) set-ups comprised of a mass body, spring, and a damper are listed below. Applying the derived solutions (Table 3.5) for the mass–spring system, a 2nd order linear homogeneous ODE, investigate the behavior of the system by conducting the following:

A. Analyze the body's behavior by plotting the change in the mass body's position (displacement, X) with time (t) given an initial displacement of 1 m ($X_{t=0} = X_0 = 1.0$, Dirichlet) and no initial velocity ($X'_{t=0} = X'_0 = 0.0$, Neumann). Note that the direction of X is downwards from the equilibrium position.

B. Under the same initial condition (Dirichlet and Neumann conditions) as previously stated, examine the time-variation of the mass body's position against its velocity (X') by plotting a phase diagram with X-axis describing the body's velocity and Y-axis describing its position (phase diagram).

Table 3.4 Mass–spring set-up

Set-up	Mass, m [kg]	Spring constant, k [kg/s^2]	Resistance coefficient c [kg/s]
(1)	1.0	1.0	0.0
(2)	1.0	1.0	1.0
(3)	1.0	1.0	2.0
(4)	1.0	1.0	3.0

Table 3.5 Summary of solutions for eq. 3.16 and corresponding physical phenomena

Solution type	Form of Solution	Physical Phenomena
Imaginary $(c_p = 0)$	$X = X_0 \cos \omega_p t$ $(\lambda = \pm i\omega_p)$	Natural vibration
Imaginary $(c_p < 2\omega_p)$	$X = X_0 e^{pt}(\cos qt - (p/q)\sin qt)$ $(\lambda = p \pm iq)$	Damped: Underdamping
Double root $(c_p = 2\omega_p)$	$X = X_0(1 - \alpha t)e^{\alpha t}$ $(\lambda = \alpha)$	Damped: Critical damping
Distinct roots $(c_p > 2\omega_p)$	$X = X_0(\alpha - \beta)^{-1}(-\beta e^{\alpha t} + \alpha e^{\beta t})$ $(\lambda = \alpha, \beta)$	Damped: Overdamping

3.6.2 Procedures for solving Exercise 2

Here are the steps to constructing the time-series and phase diagram. Import the necessary modules (see previous exercise).

```
1   In [1]: import matplotlib.pyplot as plt
2           import numpy as np
```

Here we construct user-defined functions to avoid copying and pasting every time. Using the syntax "def function_name(arguments)," the following functions to represent the various general solutions of the mass–spring system are derived. The arguments and the variables are locally stored in the function. In other words, the variable names you set will not be used outside of the function. Values X and V (separated by comma) are then returned to the position in the program where the function was called. All function-related codes are indented below the function.

The first function is for the undamped condition (harmonic oscillation).

```
1   In [2]: def nat(t,Xinit,cp,wp):
2               X = Xinit*np.cos(wp*t)
3               V = -wp*Xinit*np.sin(wp*t)
4               return X,V
```

The second function is for the underdamped condition.

```
1   In [3]: def under_dam(t,Xinit,cp,wp):
2               p = -cp/2.
3               q = np.sqrt(4*wp*wp-cp*cp)/2.
```

```
4                    X = Xinit*np.exp(p*t)*(np.cos(q*t)-(p/
                         q)*np.sin(q*t))
5                    V = -Xinit*np.exp(p*t)*(q+((p*p)/q))*
                         np.sin(q*t)
6                    return X,V
```

The third function is for the critically damped condition.

```
1   In [4]: def crit_dam(t,Xinit,cp,wp):
2                    a = -cp/2.
3                    X = Xinit*(1-a*t)*np.exp(a*t)
4                    V = -Xinit*np.power(a,2)*t*np.exp(a*t)
5                    return X,V
```

The fourth function is for the overdamped condition.

```
1   In [5]: def over_dam(t,Xinit,cp,wp):
2                    a = (-cp+np.sqrt(cp*cp-4*wp*wp))/2.
3                    b = (-cp-np.sqrt(cp*cp-4*wp*wp))/2.
4                    X = Xinit*(-b*np.exp(a*t)+a*np.exp(b*t
                         ))/(a-b)
5                    V = a*b*Xinit*(np.exp(b*t)-np.exp(a*t)
                         )/(a-b)
6                    return X,V
```

After constructing the user-defined functions, we can construct the main code similar to the previous exercise.

We first define the time coordinate and the initial conditions.

```
1   In [6]: #Setting up the time coordinate
2                    t = np.arange(0,15.1,0.1)
3
4                    #Initial conditions
5                    Xinit = 1.
```

Then we input the parameters described in the Table 3.4. Let us start with the first set-up (case 1)

```
1   In [7]: ###Case 1
2                    #Mass Set-up
3                    m = 1.0
4                    k = 1.0
5                    c = 0.0
6                    cp = c/m
7                    wp = np.sqrt(k/m) #np.sqrt is a square
                         root function
```

Here, another concept called conditional statement is used. Conditional statements are essential and common to programming. It is useful for us to decide which conditions the mass–spring system behaves depending on the internal parameters c_p and ω_p. Similar to the default condition, the codes that meet the condition stated are indented directly below the if function. 'elif' is used if more conditions are necessary.

```
1   In [8]: if cp==0.:
2               X,V=nat(t,Xinit,cp,wp)
3           elif cp<2*wp:
4               X,V=under_dam(t,Xinit,cp,wp)
5           elif cp==2*wp:
6               X,V=crit_dam(t,Xinit,cp,wp)
7           else:
8               X,V=over_dam(t,Xinit,cp,wp)
```

In order to compare the behavior with other cases, let us store the X and V to variables X1 and V1 for Case 1.

```
1   In [9]: X1=X
2           V1=V
```

Correspondingly, the other cases are constructed in the same manner.

```
1   In [10]: ###Case 2
2            #Mass Set-up
3            m = 1.0
4            k = 1.0
5            c = 1.0
6            cp = c/m
7            wp = np.sqrt(k/m)
8
9            if cp==0.:
10               X,V=nat(t,Xinit,cp,wp)
11           elif cp<2*wp:
12               X,V=under_dam(t,Xinit,cp,wp)
13           elif cp==2*wp:
14               X,V=crit_dam(t,Xinit,cp,wp)
15           else:
16               X,V=over_dam(t,Xinit,cp,wp)
17           X2=X
18           V2=V
19
20           ###Case 3
21           #Mass Set-up
22           m = 1.0
23           k = 1.0
24           c = 2.0
25           cp = c/m
26           wp = np.sqrt(k/m)
27
28           if cp==0.:
29               X,V=nat(t,Xinit,cp,wp)
30           elif cp<2*wp:
31               X,V=under_dam(t,Xinit,cp,wp)
32           elif cp==2*wp:
33               X,V=crit_dam(t,Xinit,cp,wp)
```

```
34              else:
35                  X,V=over_dam(t,Xinit,cp,wp)
36          X3=X
37          V3=V
38
39          ###Case 4
40          #Mass Set-up
41          m = 1.0
42          k = 1.0
43          c = 3.0
44          cp = c/m
45          wp = np.sqrt(k/m)
46
47          if cp==0.:
48              X,V=nat(t,Xinit,cp,wp)
49          elif cp<2*wp:
50              X,V=under_dam(t,Xinit,cp,wp)
51          elif cp==2*wp:
52              X,V=crit_dam(t,Xinit,cp,wp)
53          else:
54              X,V=over_dam(t,Xinit,cp,wp)
55
56          X4=X
57          V4=V
```

Finally, we now have stored the value of displacement and velocity as functions of time for all cases. It is now time to plot them. Below is again a more specific syntax to plotting. Since we are required two plots, time-series and phase diagram, we can construct a plot that contains both of them side-by-side.

```
1   In [11]: f,(ax1,ax2)=plt.subplots(1,2,figsize=(11,
        5)) #plt.subplots(no. of rows, no. of columns,
        figsize=(size
```

Once the axes of the figures are created, we can fill each axes through the following codes.

```
1   In [12]: ax1.plot(t,X1,'r',label='Case 1') #'r'
        refers to the color red. 'label' is an attribute
        ax1.plot(t,X2,'b',label='Case 2') #'b' refers to
        the color blue.
2           ax1.plot(t,X3,'g',label='Case 3') #'g'
                refers to the green color.
3           ax1.plot(t,X4,'k',label='Case 4') #'k'
                refers to the black color.
4           ax1.legend(loc=4) #To plot the legend at
                quadrant 4
5           ax1.set_title('Time Series Plot')
6           ax1.set_xlabel('Time (s)')
7           ax1.set_ylabel('Displacement (m)')
8           ax1.grid(True) #Includes grid lines
9
10          ax2.plot(V1,X1,'r',label='Case 1')
11          ax2.plot(V2,X2,'b',label='Case 2')
12          ax2.plot(V3,X3,'g',label='Case 3')
```

```
13            ax2.plot(V4,X4,'k',label='Case␣4')
14            ax2.legend(loc=4)
15            ax2.set_title('Phase␣Diagram')
16            ax2.set_xlabel('Velocity␣(m/s)')
17            ax2.grid(True)
```

Once the plot is ready, we can show,

```
1  In [13]: plt.show()
```

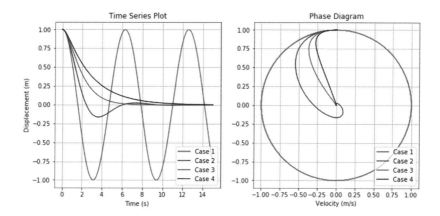

It can be seen that the case of undamped condition has a period of 2π s. or 6.28 s. From the time-series plot, in the case of no resistance (undamped, case 1), the natural vibration of the system is exhibited (harmonic oscillation). Upon introducing a damper of resistance coefficient 1.0 kg/s, its angular frequency is slightly smaller than the undamped condition until the system finally attenuates. When the resistance coefficient is increased to 2.0 kg/s, the displacement recedes and immediately returns to the equilibrium position. When the resistance coefficient is increased further to 3.0 kg/s, it becomes an overdamped solution thus taking more time for the displacement to attenuate. From the phase diagram, it can be shown that the phase of the harmonic oscillation is orbiting infinitely. The phase of all damped conditions converge to the origin. The undamped condition manifests a spiral movement owing to the passing of the object above the equilibrium condition. The critically damped and the overdamped condition show similar phase curves but it can be confirmed that throughout, the velocity of the critical damping condition case has a larger upward velocity (negative velocity) at the same displacement than the overdamped condition case. This is consistent with the fact that under critical damping, the displacement attenuates the fastest.

⬥⬥⬥⬥⬥⬥ Problems for Exercise ⬥⬥⬥⬥⬥⬥

3.1 Find the exact solution of the 2nd order linear homogeneous ODE shown below. The internal parameters are assumed constant.

1. $T'' - 4T' + 4T = 0$, $T(0) = 3$, $T'(0) = 4$

2. $T'' - 3T' + 2T = 0$, $T(0) = 2$, $T'(0) = 0$

3. $T'' + 4T' + 5T = 0$, $T(0) = 2$, $T'(0) = 0$

4. $T'' + 6T' + 9T = 0$, $T(0) = -1$, $T'(0) = 3$

5. $T'' - T' - 6T = 0$, $T(0) = 1$, $T'(0) = 2$

6. $T'' + 2T' + 5T = 0$, $T(0) = 1$, $T'(0) = -4$

3.2   Since the 2nd order ODE shown below are non-homogeneous or non-linear, the solution process introduced in this chapter will not be applicable as it is. However, converting$^{\circ}$**109** them to 1st-order ODEs, a general solution can be derived.

$\circ$**109** When encountering a difficult problem, try to simplify the given problem into a form that is solvable. This is a conventional method to scientific understanding.

1. $\frac{d^2T}{dt^2} = \frac{2}{t}$ $^{\circ}$**110**

$\circ$**110** Can be directly integrated with respect to t.

2. $\frac{d^2T}{dt^2} - [\frac{dT}{dt}]^2 + 9 = 0$ $^{\circ}$**111**

$\circ$**111** Set dT/dt to X.

3. $\frac{1}{2t}\frac{d^2T}{dt^2} - [\frac{dT}{dt}]^2 = 0$ $^{\circ}$**112**

$\circ$**112** Set dT/dt to X.

3.3   As shown in the figure below, a pendulum with a mass m [kg] is attached to a thread of length L [m]. The angle of displacement from the vertical axis is assumed to be θ. Given that θ is sufficiently small, $\sin\theta \cong \theta$. Answer the following questions.

1. Derive the equation of motion of the pendulum and find its natural angular frequency$^{\circ}$**113**.

$\circ$**113** The result should be an ODE similar to the mass-point system neglecting resistance.

2. Consider the two identical pendulums are placed on the earth and the moon, What is the ratio of the natural angular frequencies of both pendulums?

3. Compare the difference in behavior between the pendulum introduced in this problem and the mass–spring system introduced in ch.3°**114**?

$\circ$**114** Is gravity an internal parameter or an external parameter?

4. Derive the law of energy conservation of the pendulum's motion by integrating the pendulum's equation with the distance traveled along the arc.

5. Establish an energy conservation law of the pendulum's motion, and

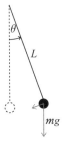

differentiate it in order to derive the equation of motion of the pendulum.

4

2nd order linear non-homogeneous ODE (resonance phenomena)

Using the mass–spring system experiencing resistance and the RCL electric circuit discussed in the previous chapter, we will investigate the system when both systems experience a periodic external force. The periodic external forces correspond to the non-homogeneous term (constants, independent variable terms), and these systems are represented by 2nd order linear non-homogeneous ODE, the subject of this chapter. A general solution of the non-homogeneous form is obtained by adding a particular solution of non-homogeneous form to the general solution of the homogeneous form explained in the previous chapter. Due to the action of the periodic external force, the behavior of the system becomes diverse. Basically, it is a superposition of solutions, one that depends on the natural angular frequency (internal system), and another that depends on the angular frequency of the external system. An interesting phenomenon which will be introduced is "resonance." When the natural angular frequency and the external force's angular frequency are the same and the resistance becomes negligible, the amplitude increases with time, which will lead to collapse or failure of the system. Also, when the natural angular frequency and the external forces' angular frequency are almost equal but remain different from each other, a phenomenon called "beats" occur.

○115   As in the previous chapter, the internal parameters are also assumed constant in this chapter. Although it is possible to derive an exact solution for non-constant internal parameters, the process is not easy. In such cases, numerical methods to be introduced in the next chapter is recommended.

○116   The external system differs from the internal system in terms of its relationship with the dependent variable. Terms do not contain the dependent variable in the external system. Rather, they are comprised of the independent variable or a constant. In the previous chapter, the gravitational force in the mass–spring system and the DC voltage supply in the RCP electric circuit seem to comprise the external system but of no influence to the system's total behavior.

4.1   Deriving the exact solution

If the dependent variable is T, the independent variable is t, the internal parameter constants are a and b○115, and the external system represented by the function $g(t)$, the 2nd order linear homogeneous ODE and the linear non-homogeneous ODE can be generalized by the following equations. Physically, both ODE types differ in terms of the work done by an external system○116.

2nd order linear homogeneous ODE $\dfrac{d^2 T}{dt^2} + a\dfrac{dT}{dt} + bT = 0.$ (4.1)

2nd order linear non-homogeneous ODE

$$\frac{d^2 T}{dt^2} + a\frac{dT}{dt} + bT = g(t).$$ (4.2)

(Internal system)   (External system)

The general solution of eq. 4.1 can be expressed by a linear combination of two linearly independent solutions (basis of solutions) T_1, T_2 to eq. 4.1 (superposition principle).

$$T = C_1 T_1 + C_2 T_2 \qquad (4.3)$$

Here, C_1 and C_2 are undetermined constants which can be determined from the initial or boundary conditions. The method of finding the solution of this linear homogeneous ODE has already been discussed in detail in the previous chapter. The solution of the linear non-homogeneous ODE can be expressed as follows[117] by first obtaining a general solution of a linear homogeneous ODE and adding a particular (special) solution T_s which satisfies eq. 4.2.

$$T = C_1 T_1 + C_2 T_2 \qquad + \quad T_s \qquad (4.4)$$

| General solution | = General solution | + Particular solution |
| of eq. 4.2 | of eq. 4.1 | of eq. 4.2 |

The particular solution T_s is special in the sense that it must not include an undetermined constant yet satisfies as a solution to the linear non-homogeneous ODE (eq. 4.2). T_s depends on the functional form of $g(t)$, which represents the external system, and the solution type (distinct real roots, imaginary roots, or double real roots) of the characteristic equation of the lineary homogeneous ODE (eq. 4.1). The functional form of $T_s(t)$ is organized as shown in Table 4.1.

Table 4.1 Functional forms of the particular solution T_s to eq. 4.2

	Functional form of $g(t)$ in eq. 4.2	Characteristic equation $(\lambda^2 + a\lambda + b = 0)$ of eq. 4.1	Particular solution of eq. 4.2
Ia		(1) $\lambda \neq 0$	(1) $P_n(t)$
Ib	$p_n(t)$[118]	(2) $\lambda = 0$ distinct roots	(2) $tP_n(t)$
Ic		(3) $\lambda = 0$ double root	(3) $t^2 P_n(t)$
IIa		(1) $\lambda \neq \alpha$	(1) $P_n(t)e^{\alpha t}$
IIb	$p_n(t)e^{\alpha t}$[119]	(2) $\lambda = \alpha$ distinct roots	(2) $tP_n(t)e^{\alpha t}$
IIc		(3) $\lambda = \alpha$ double root	(3) $t^2 P_n(t)e^{\alpha t}$
IIIa	$p_n(t)\sin\beta t$ or[120]	(1) $\lambda \neq \beta i$	(1) $P_n(t)\sin\beta t$ $+Q_n(t)\cos\beta t$
IIIb	$p_n(t)\cos\beta t$	(2) $\lambda = \beta i$	(2) $t\{P_n(t)\sin\beta t$ $+Q_n(t)\cos\beta t\}$

$p_n(t), P_n(t)$, and $Q_n(t)$ are polynomials of order n and are expressed by the following. The degree n of the polynomial should match $p_n(t)$[121].

$$g(t) : p_n(t) = D_n t^n + D_{n-1} t^{n-1} + \cdots\cdots + D_1 t + D_0$$
$$g(t) : P_n(t) = A_n t^n + A_{n-1} t^{n-1} + \cdots\cdots + A_1 t + A_0$$
$$Q_n(t) = B_n t^n + B_{n-1} t^{n-1} + \cdots\cdots + B_1 t + B_0$$

The coefficients of the polynomial functions shown in Table 4.1 can be solved by substituting T_s into eq. 4.2. Once the special solution T_s is solved, the undetermined constants C_1 and C_2 are then determined using the initial and boundary conditions similar to the procedure conducted in the previous chapter. The process of solving for the exact solution of a 1st order linear non-homogeneous ODE is summarized in Figure 4.1. This procedure is also called "the Method of Undetermined Coefficients."

[117] The process of "superposition principle" exhaustively used in the previous chapters is strictly applicable to linear homogeneous ODEs. The same process will not apply for linear non-homogeneous expressions which include constants and independent variable terms. Instead, a solution which involves the "superposition principle" is constructed which is a sum of the "solution of a linear homogeneous equation derived by setting the external terms to zero" (eq. 4.1) and a "particular solution" (eq. 4.2). Linearity is an important factor to derive a good solution. These methods will not apply for non-linear cases.

[118] Group I are for ODEs with external functions expressed by polynomials of independent variables. Regardless of the parameters of the external function or when the solution of the characteristic equation is zero, various cases may be decided to derive the special solution.

[119] Group II is when the external function includes the *exp* function. Various cases depending on the condition of the external parameter may be selected. Each case has a different effect to the particular solution. The external parameter contained in the *exp* term may introduce exponential decay or increase in the dependent variable of the system.

[120] Group III is when the external function includes trigonometric functions. Its frequency, which is an external parameter, affects the particular solution. Cases are available depending on the value of frequency.

[121] This means to assume that the particular solution will be a polynomial of similar form as the external function. Depending on the form of the solution of the characteristic equation and the external parameters, the polynomial may be multiplied by a function of the independent variable or t^2.

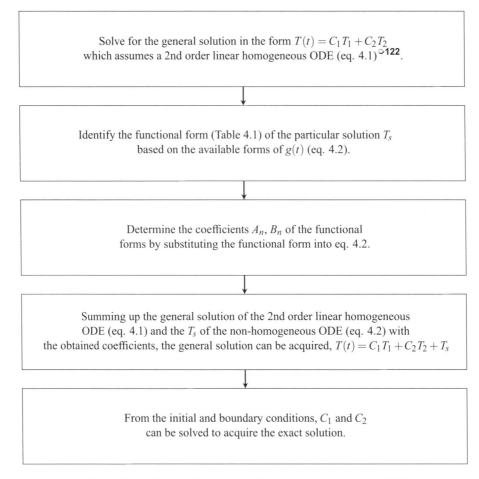

Solve for the general solution in the form $T(t) = C_1 T_1 + C_2 T_2$
which assumes a 2nd order linear homogeneous ODE (eq. 4.1)[$^{\circlearrowright}$**122**].

Identify the functional form (Table 4.1) of the particular solution T_s
based on the available forms of $g(t)$ (eq. 4.2).

Determine the coefficients A_n, B_n of the functional
forms by substituting the functional form into eq. 4.2.

Summing up the general solution of the 2nd order linear homogeneous
ODE (eq. 4.1) and the T_s of the non-homogeneous ODE (eq. 4.2) with
the obtained coefficients, the general solution can be acquired, $T(t) = C_1 T_1 + C_2 T_2 + T_s$

From the initial and boundary conditions, C_1 and C_2
can be solved to acquire the exact solution.

Figure 4.1   Flowchart to solving a linear non-homogeneous ODE

$^{\circlearrowright}$**122**   Finally, a particular solution is to be added to the general solution of the linear homogeneous ODE. Undetermined constants, C_1 and C_2, should not be solved in this step.

4.2   Behavior of mass–spring system with periodic external force

Consider an internal system where a body (mass point) of mass m is connected to a spring with a spring constant k and a damper of resistance coefficient c similar to Figure 3.1. This time, we introduce an external system which exerts an external force that varies by a sine function of time with an angular frequency ω and an amplitude F_0 as shown[$^{\circlearrowright}$**123**].

$^{\circlearrowright}$**123**   The unit of each physical property must ensure consistency of the terms in eq. 4.7.
Mass [kg]   Spring constant [kg/s²]
Resistance coefficient [kg/s]   Angular frequency [1/s]
Amplitude [kg-m/s²]
Since trigonmetric functions are dimensionless, the amplitude of the external force has the same dimension or units as the inertial force.

$$g(t) = F_0 \sin \omega t \tag{4.5}$$

The mass point shall be in a static equilibrium state as the initial condition. Its movement shall be caused by the action of the external force (eq. 4.5). Assuming the downward displacement from the equilibrium position is X and the balance of the forces in the static equilibrium state represented by eq. 4.6, the inertial force, gravitation, the spring's restoring force, air (damping) resistance, is in equilibrium with the external force at any displacement X (eq. 4.7)

$$mg = ks \tag{4.6}$$

$$m\frac{d^2X}{dt^2} = mg - k(s+X) - c\frac{dX}{dt} + F_0 \sin \omega t \tag{4.7}$$

Substituting eq. 4.6 into eq. 4.7 results in a 2nd order linear non-homogeneous ODE with displacement X as the dependent variable, and t as the independent variable.

$$m\frac{d^2X}{dt^2} + c\frac{dX}{dt} + kX = F_0 \sin \omega t \tag{4.8}$$

(Internal system)   (External system)

From eq. 4.8, the gravity term disappears and the system is composed of three internal systems (inertial force, resistance, restoring force of the spring) and an external system. Here, the internal parameters are 3 (mass m, spring constant k, and resistance coefficient c), while the external parameters are 2 (the external force's angular frequency ω, and the amplitude F_0). By dividing all terms by m, the coefficient of the second derivative vanishes reducing the internal parameters into 2 ($c_p = c/m, \omega_p = \sqrt{k/m}$), while the external parameters remain 2 ($F_p = F_0/m, \omega$).

$$\frac{d^2X}{dt^2} + \left(\frac{c}{m}\right)\frac{dX}{dt} + \left(\frac{k}{m}\right)X = \left(\frac{F_0}{m}\right)\sin \omega t$$

$$\frac{d^2X}{dt^2} + c_p\frac{dX}{dt} + \omega_p^2X = F_p \sin \omega t \tag{4.9}$$

Since the combination of the two internal system parameters (c_p, ω_p) and the behavior of the solution are described in detail in the previous chapter, we shall assume that the resistance c_p is equal to 0. We will investigate the system based on the relationship of ω_p and the external parameter. The system is simplified by the absence of damping resistance and it becomes as shown in eq. 4.10.

$$\frac{d^2X}{dt^2} + \omega_p^2X = F_p \sin \omega t \tag{4.10}$$

(Internal system)   (External system)

The physical units of each term in eq. 4.10 can be inspected. Coinciding with the physical unit of the 1st term on the left-hand side (m/s^2), the units of ω_p is 1/s and F_p is m/s^2. The angular frequency ω of the external force and the natural angular frequency ω_p of the internal system have the same physical unit. Following the instructions in Figure 4.1, let us find the solution of eq. 4.10.

4.2.1 Deriving a general solution of linear homogeneous ODE

Given that the external force in the right-hand side of eq. 4.10 is 0, the

⟳**124** $\lambda^2 + \omega_p^2 = 0$

solution of the characteristic equation⟳**124** is

$$\lambda = \pm i\omega_p \tag{4.11}$$

Here, the characteristic equation is imaginary with $p = 0$ and $q = \omega_p$ (Table 3.1). The general solution of the ODE becomes,

$$X = C_1 \cos \omega_p t + C_2 \sin \omega_p t \tag{4.12}$$

4.2.2 Particular solution of linear non-homogeneous ODE

Let us refer the external force in eq. 4.5 to Table 4.1. Since the external system divided by mass is $F_p \sin \omega t$, it corresponds to the case III: $p_n(t) \sin \beta t$ in the table. The polynomial p_n is the constant F_p, with an order n of zero order (i.e. it does not depend on t), and $\sin \beta t$ is $\sin \omega t$. From here, the classification of the particular solution $T_s(t)$, will depend on whether the angular frequency ω is equivalent with the internal system's natural angular frequency ω_p⟳**125**.

⟳**125** The relationship of the natural angular frequency of the internal system and the angular frequency of the external system influences the particular solution.

◆ **When external angular frequency ω and natural angular frequency ω_p are unequal ($\omega \neq \omega_p$)**

Since the order of $P_n(t)$ and $Q_n(t)$ are the same as $p_n(4)$ which is 0 as shown in IIIa of Table 4.1, the particular solution X_s can be written temporarily as follows,

$$X_s = A_0 \sin \omega t + B_0 \cos \omega t \tag{4.13}$$

Substituting the particular solution X_s above (eq. 4.13) into the linear non-homogeneous ODE (eq. 4.10)⟳**126**,

⟳**126**
$X_s' = \omega A_0 \cos \omega t - \omega B_0 \sin \omega t$
$X_s'' = -\omega^2 A_0 \sin wt - \omega^2 B_0 \cos \omega t$
Substituting the above to eq. 4.10,
$[-\omega^2 A_0 \sin \omega t - \omega^2 B_0 \cos \omega t]$
$+\omega_p^2[A_0 \sin \omega t + B_0 \cos \omega t]$
$= F_p \sin \omega t$

$$(\omega_p^2 A_0 - \omega^2 A_0 - F_p) \sin \omega t + (\omega_p^2 B_0 - \omega^2 B_0) \cos \omega t = 0$$

Since the above equation must always hold with respect to any arbitrary independent variable t, likewise, it must always be satisfied with any arbitrary $\sin \omega t$ and $\cos \omega t$. Therefore, A_0 and B_0 must be equal to the following for the condition $\omega \neq \omega_p$.

$$A_0 = \frac{F_p}{-\omega^2 + \omega_p^2}, \quad B_0 = 0$$

From the above, the particular solution can be obtained as follows,

$$X_s = \frac{F_p}{\omega_p^2 - \omega^2} \sin \omega t \tag{4.14}$$

◆ **When external angular frequency ω and natural angular frequency ω_p are equal ($\omega = \omega_p$)**

Here, we refer to eq. IIIb of Table 4.1 because $\lambda = \pm i\omega$ and ω is equivalent to the β of he functional form $p_n(t) \sin \beta t$. Since the order of $P_n(t)$ and $Q_n(t)$ is the same as $p_n(t)$ which is 0, the particular solution X_s can be written as follows,

$$X_s = A_0 t \sin \omega_p t + B_0 t \cos \omega_p t \tag{4.15}$$

Substituting the particular solution X_s above (eq. 4.15) into the linear non-homogeneous ODE (eq. 4.10)[127],

$$(-F_p - 2\omega_p B_0)\sin\omega_p t + 2\omega_p A_0\cos\omega_p t = 0.$$

The above equation must always hold with respect to any arbitrary independent variable t; likewise, it must always be satisfied with any arbitrary $\sin\omega t$ and $\cos\omega t$. Therefore, A_0 and B_0 must be equal to the following,

$$A_0 = 0, \quad B_0 = \frac{-F_p}{2\omega_p} \tag{4.16}$$

From the above, the particular solution is as follows,

$$X_s = -\frac{F_p}{2\omega_p}t\cos\omega_p t \tag{4.17}$$

4.2.3 Deriving an exact solution of linear non-homogeneous ODE

Up to this point, we have obtained a general solution of the linear homogeneous ODE (ignoring the influence of external forces), and a particular solution of the linear non-homogeneous ODE (considering the external force). By adding both equations, the final goal is to find the exact solution of the linear non-homogeneous ODE. To do that, the undetermined constants must be determined by satisfying the initial condition.

◆ **When external angular frequency ω and natural angular frequency ω_p are not equal ($\omega \neq \omega_p$): 2 periodic solutions and beats**

The general solution (eq. 4.18) of the linear non-homogeneous ODE is obtained by adding the general solution (eq. 4.12) of the linear homogeneous ODE and the particular solution (eq. 4.14),

$$X = C_1\sin\omega_p t + C_2\cos\omega_p t + \frac{F_p}{\omega_p^2 - \omega^2}\sin\omega t \tag{4.18}$$

Given the initial condition ($t = 0$) of $X = 0$ (equilibrium state) and $X' = 0$ (stationary state), the undetermined constants will then be,

When $t = 0, X = 0, C_2 = 0$.

When $t = 0, X' = 0, C_1 = \dfrac{-F_p(\omega/\omega_p)}{\omega_p^2 - \omega^2}$ [128]

Substituting the constants into eq. 4.18, an exact solution of linear non-homogeneous ODE was found,

$$X = \frac{F_p}{\omega_p^2 - \omega^2}\left(-\frac{\omega}{\omega_p}\sin\omega_p t + \sin\omega t\right) \tag{4.19}$$

$$X' = \frac{F_p\omega}{\omega_p^2 - \omega^2}\left(-\cos\omega_p t + \cos\omega t\right)$$

Let us interpret the behavior based on the solution. Sum of two sine

[127] $X_s' = A_0\sin\omega_p t + B_0\cos\omega_p t + \omega_p A_0 t\cos\omega_p t - \omega_p B_0 t\sin\omega_p t$
$X_s'' = \omega_p A_0\cos\omega_p t - \omega_p B_0\sin\omega_p t + \omega_p A_0\cos\omega_p t - \omega_p B_0\sin\omega_p t - \omega_p^2 A_0 t\sin\omega_p t - \omega_p^2 B_0 t\cos\omega_p t$
$= 2\omega_p(A_0\cos\omega_p t - B_0\sin\omega_p t) - \omega_p^2 X_s$
Substituting the above to eq. 4.10,
$2\omega_p(A_0\cos\omega_p t - B_0\sin\omega_p t) - \omega_p^2 X_s + \omega_p^2 X_s = F_p\sin\omega t$

[128]
$X' = C_1\omega_p\cos\omega_p t - C_2\omega_p\sin\omega_p t + (\omega F_p/(\omega_p^2 - \omega^2))\cos\omega t$

functions can be seen. One has a natural angular frequency of the internal system, and the other has the angular frequency of the external system. The output which arises from two periodic functions of different angular frequencies is called **superposition of two harmonic oscillations**. The fact that the amplitude of the fluctuation is proportional to the external system's amplitude F_p can be understood intuitively. What is interesting is that the amplitude of the fluctuation also depends on the magnitude of the relationship between the 2 angular velocities ω and ω_p. Let us investigate this further.

① $\omega \ll \omega_p$: When the angular frequency of the external system ω is sufficiently smaller than the natural angular frequency ω_p of the system, the time variation (frequency, period) is determined by the angular frequency of the external system; and the amplitude of the displacement is determined$^{◌129}$ by the amplitude F_p of the external system and the natural angular frequency ω_p of the internal system.

$$X \cong \frac{F_p}{\omega_p^2} \sin \omega t \qquad (4.20)$$

② $\omega \gg \omega_p$: When the angular frequency of the external system is sufficiently large compared with the natural angular frequency of the internal system, the pattern of the displacement is determined by the natural angular frequency of the internal system. Meanwhile, the amplitude of the displacement is determined by the amplitude of the external system, the natural angular vibration, and the external system's angular frequency$^{◌130}$.

$$X \cong \frac{F_p}{\omega_p \omega} \sin \omega_p t \qquad (4.21)$$

In both cases ① ② , the over-all pattern of the displacement in terms of frequency is governed by the smaller angular frequency; that is, a slow variation (long cycle). With a neatly constant value, the amplitude is determined by the external force and the angular frequencies.

③ $\omega \cong \omega_p$: Let us calculate an approximate solution wherein both ω and ω_p have almost negligible difference as represented by $\varepsilon = \omega_p - \omega$ $(\omega/\omega_p \cong 1, \omega + \omega_p \cong 2\omega_p)$.

$$
\begin{aligned}
X &= \frac{F_p}{(\omega_p - \omega)(\omega_p + \omega)} \left(-\frac{\omega}{\omega_p} \sin \omega_p t + \sin \omega t \right) \quad ^{◌131} \\
&\cong \frac{F_p}{2\varepsilon\omega_p} (-\sin \omega_p t + \sin \omega t) \\
&\cong \frac{F_p}{2\varepsilon\omega_p} \left[2\cos\left(\frac{\omega + \omega_p}{2} t \right) \sin\left(\frac{\omega - \omega_p}{2} t \right) \right] \\
&\cong \frac{-F_p}{\varepsilon\omega_p} \left\{ \cos(\omega_p t) \sin\left(\frac{\varepsilon}{2} t \right) \right\} \\
&\cong \left\{ \frac{-F_p}{\varepsilon\omega_p} \sin\left(\frac{\varepsilon}{2} t \right) \right\} \cos \omega_p t \qquad (4.22)
\end{aligned}
$$

The time variation of the displacement (frequency, period) seems to be

Margin notes:

◌**129** $\omega_p^2 - \omega^2 = \omega_p^2(1 - \omega^2/\omega_p^2)$
$\cong \omega_p^2,$
$\omega/\omega_p \cong 0$

◌**130** $[F_p/(\omega_p^2 - \omega^2)][-(\omega/\omega_p)$
$\sin \omega_p t + \sin \omega t],$
$\cong \{F_p/[\omega^2(\omega_p^2/\omega^2 - 1)]\}[-(\omega/\omega_p)$
$\sin \omega_p t], \cong (F_p/\omega\omega_p) \sin \omega_p t$

◌**131** Utilize the trigonometric property,
$\sin A - \sin B =$
$2\cos((A+B)/2)\sin((A-B)/2)$

determined by the frequency $\omega \cong \omega_p$ as shown in $\cos \omega_p t$. However, the amplitude is no longer constant but also varying periodically. It should be noted here that the angular frequency $\varepsilon/2$ of the periodic fluctuation of the amplitude is much smaller (i.e. a long cycle) than the original angular frequency $\omega \cong \omega_p$ of the system. Moreover, its maximum amplitude $F_p/\varepsilon \omega_p$ can be larger than the amplitude of ① ② because the miniscule value ε is in the denominator. At very close values of frequencies, two functions exist in one solution wherein one function manifests a slow variation of a very large amplitude. This type of oscillation is called **beats**.

Tuning of guitars by a tuning fork is an example of beats. When pressing the tuning fork (external system) on the body of the guitar, it produces a sound of certain frequency brought about by strumming the guitar strings (internal system). When the natural angular frequency (determined by the string type and tension) of the string of the guitar is slightly varied from the tuning fork to be match, a totally different sound which varies in frequency to the tuning fork and the guitar is heard. Owing to this additional frequency exhibited when the frequencies of the guitar string and the tuning fork are matched, the guitar can be tuned more accurately. If there is a wave tank in the laboratory of the university, it is also possible to create a beat from the water waves. If two sine waves with the same amplitude and slightly different frequencies are created in the wave tank, another wave of large amplitude fluctuations differing from the individual waves appear. It is interesting that even if the phenomenon is linear, a peculiar phenomenon is manifested[132].

◆ **When external angular frequency ω and natural angular frequency ω_p are equal ($\omega = \omega_p$): Resonance**

The general solution (eq. 4.23) of the linear non-homogeneous ODE is obtained by adding the general solution (eq. 4.12) of the linear homogeneous ODE and the particular solution (eq. 4.17) of the linear non-homogeneous equation.

$$X = C_1 \sin \omega_p t + C_2 \cos \omega_p t - \frac{F_p}{2\omega_p} t \cos \omega_p t \qquad (4.23)$$

Given the initial condition ($t = 0$) of $X = 0$ (equilibrium state) and $X' = 0$ (stationary state), the undetermined constants will then be,

When $t = 0, X = 0, C_2 = 0$.

When $t = 0, X' = 0, C_1 = \dfrac{F_p}{2\omega_p^2}$ [133]

From the above, the exact solution of the linear non-homogeneous ODE was found.

$$X = \frac{F_p}{2\omega_p^2} \left(\sin \omega_p t - t\omega_p \cos \omega_p t \right) \qquad (4.24)$$

$$X' = \frac{F_p}{2} \left(t \sin \omega_p t \right) \qquad (4.25)$$

Let us interpret the behavior based on the exact solution above[134].

○**132** It seems that a new sinusoidal wave with a period completely different from the natural and external vibrations is produced as shown in eq. 4.22. In fact, "beats" is observable visually (wave) and auditorially (tune). However, as mentioned in ch. 1, exact solutions of linear systems do not deviate from the general solution (i.e. frequency of waves should not change). Beats is a phenomon generated from superposition of sine waves of close frequencies. If beats are decomposed into waves of varying frequencies, two sine waves appear as before it was approximated. This phenomenon is unique and should not be mistaken for a non-linear phenomenon (ch. 6).

○**133** $X' = C_1 \omega_p \cos \omega_p t - C_2 \omega_p \sin \omega_p t - (F_p/2\omega_p) \cos \omega_p t + (F_p/2) t \sin \omega_p t$

○**134** Observe carefully the equation which represents the exact solution. Unlike approximate solutions, the expected behavior of the system can be guessed from the equation representing the exact solution.

Comparing eq. 4.24 with eq. 4.19 brought about by the conditions of the frequencies of the internal and external system, significant differences in the solution can be found. Particularly, the presence of the second-term of the right-hand side $t\,\omega_p \cos \omega_p t$ is comprised of a product of a trigonometric function and time t. It means that as t approaches ∞, the displacement also approaches infinity. This is called **resonance** phenomenon. Despite the fact that the external force is periodic, the displacement exceeds the amplitude of the external force and eventually diverges only as a result of the interaction between the internal and external system. In the real world, the mass–spring device will be destroyed[135].

⤳**135** From the perspective of law of conservation of energy, energy is balanced by natural vibration in the absence of resistance. If work is exerted incessantly by periodic external forces, energy will accumulate and diverge over time.

Many accidents are believed to be caused by this phenomenon. The most famous was the Tacoma bridge collapse (Washington State, USA) that occurred only a few months after its opening in the 1940s. The initial design of the bridge was for it to withstand a wind speed of 60 m/s, but the bridge violently vibrated even at a wind speed of 19 m/s. In spite of the low wind speed, the drag of the building created a vortex at the leeward of the bridge (Kármán vortex) which caused a period external forcing upon the bridge. The external vibration matching with the torsional vibration (its natural vibration internal system) of Tacoma bridge led to the bridge's collapse. It is from learning through these disastrous events that triggered revisions in structural designs. Owing to this, structures are now designed to prevent the occurrence of resonance. The resonance phenomenon is also applicable to earthquakes.

4.3 Behavior of current in RLC electric circuit with AC voltage

Using a similar RLC circuit as in Sect. 3.3, a similar linear non-homogeneous ODE to the mass–spring system will be derived if the power supply is replaced from DC to an AC voltage with a sinusoidal electromotive force equal to $E_0 \sin \omega t$ (Figure 4.2). The RLC circuit is represented by a resistor of resistance [γ=V/A], capacitor of capacitance [As/V], and a coil of inductance [Vs/A] connected in series [136]. When the circuit is switched on, current I will flow. The sum of the amount of voltage drop in the circuit is zero according to Kirchhoff's second voltage law. Unlike the earlier RLC circuit where the DC voltage is time-invariant, the sinusoidal electromotive force will result in an ODE with a term representing the external system (eq. 4.28).

⤳**136** The amplitude of the AC voltage is E_0 [V], and its angular frequency is [1/s].

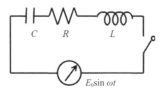

Figure 4.2 RLC Circuit

$$L\frac{dI}{dt} + RI + \frac{Q}{C} = E_0 \sin \omega t \tag{4.26}$$

Here, substituting the relationship between current and electric charge (eq. 4.27) into the equation and differentiating, eq. 4.28 can be acquired.

$$I = \frac{dQ}{dt} \tag{4.27}$$

$$L\frac{d^2I}{dt^2} + R\frac{dI}{dt} + \frac{1}{C}I = E_0 \omega \cos \omega t \tag{4.28}$$

(Internal system)   (External system)

Eq. 4.28 exhibits three internal systems represented by a capacitor, resistor, and inductor, and one external system represented by the AC voltage. Likewise, 3 internal parameters (inductance, resistance, and conductance) and 2 external parameters represented by the amplitude E_0 and its angular frequency ω. The equation can be simplified and the parameters can be reduced by eliminating the coefficient of the second derivative term. To do this, all terms are divided by L. This results in two internal parameters $c_p = R/L$, $\omega_p = \sqrt{1/LC}$, and the same number of external parameters with $E_p = E_0\omega/L$ and ω[137].

$$\frac{d^2I}{dt^2} + \left(\frac{R}{L}\right)\frac{dI}{dt} + \left(\frac{1}{L}\frac{1}{C}\right)I = E_0 \cos \omega t,$$

$$\frac{d^2I}{dt^2} + c_p\frac{dI}{dt} + \omega_p^2 I = E_p \cos \omega t \tag{4.29}$$

(Internal system)   (External system)

> **137** The units of c_p, ω_p (natural angular frequency), and E_p are [1/s], [1/s], and [A/s²], respectively. E_p has the dimension of current's acceleration.

This is exactly the same type as the equation of motion of the mass–spring system[138]. In the previous section, we have investigated the behavior of the solution with zero resistance. In this example, we will examine the behavior of the solution when the resistance is not zero.

> **138** Strictly speaking, the mass–spring system in the previous section has a different trigonometric function for the external force but this is besides the point.

4.3.1 Deriving a general solution of linear homogeneous ODE

The solution of the characteristic equation of the linear homogeneous ODE (right-hand side of eq. 4.29 set to zero) is expressed as follows (this time two internal parameters are used),

For this general solution, let us recall the previous chapter (Table 3.1).

$$\lambda = \frac{-c_p \pm \sqrt{c_p^2 - 4\omega_p^2}}{2} \tag{4.30}$$

Imaginary roots: $I = e^{pt}(C_1 \cos qt + C_2 \sin qt), p = \frac{-c_p}{2}, q = \frac{\sqrt{4\omega_p^2 - c_p^2}}{2}$

Distinct real roots: $I = C_1 e^{\alpha t} + C_2 e^{\beta t}, \quad \alpha, \beta = \frac{-c_p \pm \sqrt{c_p^2 - 4\omega_p^2}}{2}$

Double real root: $I = (C_1 + C_2 t)e^{\alpha t}, \quad \alpha = -\frac{c_p}{2}$ (Condition: $4\omega_p^2 = c_p^2$)

As we have already learned in the mass–spring system w/o external forcing, it can now be obvious that with time, eventually the movement of the mass point decays and the displacement approaches zero.

4.3.2 Particular solution of linear non-homogeneous ODE

Let us look at the external force (eq. 4.30) and look for it in Table 4.1. Since the external system function $E_p \cos \omega t$ corresponds to $p_n(t) \cos \beta t$, and the $p_n(t)$ is the constant E_p, the order n is zero, and $\cos \beta t$ is $\cos \omega t$, it belongs to case III (Table 4.1). The problem satisfies the assumption of type IIIa classification with the following functional form of the particular solution.

$$I_s = a \cos \omega t + b \sin \omega t \quad ^{\circlearrowright \textbf{139}} \tag{4.31}$$

$\circlearrowright$**139**
$I_s' = -a\omega \sin \omega t + b\omega \cos \omega t.$
$I_s'' = -a\omega^2 \cos \omega t - b\omega^2 \sin \omega t.$

Substituting 4.31 into 4.29, grouping them in terms of the trigonometric function, the following equation will be acquired,

$$\{a(\omega_p^2 - \omega^2) + c_p \omega b\} \cos \omega t + \{-a\omega c_p + (\omega_p^2 - \omega^2)b\} \sin \omega t = E_p \cos \omega t \quad ^{\circlearrowright \textbf{140}} \tag{4.32}$$

$\circlearrowright$**140** $-a\omega^2 \cos \omega t - b\omega^2 \sin \omega t$
$+c_p(-a\omega \sin \omega t + b\omega \cos \omega t)$
$+\omega_p^2(a \cos \omega t + b \sin \omega t)$
$= E_p \cos \omega t.$

From the requirement that the equation must hold for any independent variable t, the constants a and b are determined as follows [$\circlearrowright$**141**],

$\circlearrowright$**141** $(\omega_p^2 - \omega^2)a + c_p \omega b = E_p$
$-c_p \omega a + (\omega_p^2 - \omega^2)b = 0.$

$$a = \frac{E_p(\omega_p^2 - \omega^2)}{(\omega_p^2 - \omega^2)^2 + c_p^2 \omega^2}, \quad b = \frac{E_p c_p \omega}{(\omega_p^2 - \omega^2)^2 + c_p^2 \omega^2} \tag{4.33}$$

From the above, a particular solution I_s was derived.

4.3.3 Deriving a general solution of linear non-homogeneous ODE and its practical resonance

The general solution of the 2nd order linear non-homogeneous ODE (eq. 4.30), which is the final goal, is obtained by adding a general solution of the homogeneous form (internal system) (listed above) and the particular solution I_s (Figure 4.3). The behavior of the homogeneous general solution was introduced in the previous chapter. In the previous chapter, harmonic

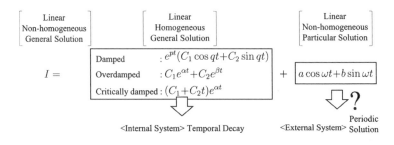

Figure 4.3 Decomposition of solution terms for linear non-homogeneous ODEs

oscillation occurs when damping is absent. Introducing dampers to the internal system will result in the displacement always returning to the zero position. On the other hand, it is obvious from eq. 4.31 that the solution does not attenuate with time. Hence, let us briefly consider the behavior of the periodic solution of non-homogeneous special solutions.

To make it easier to consider the behavior of the solution, let us synthesize the cosine and sine waves by arbitrarily relating a and b.

$$I = a\cos\omega t + b\sin\omega t = A\cos(\omega t - \eta)$$

$$\text{Here, } A = \sqrt{a^2 + b^2} \text{ and } \tan\eta = \frac{b}{a} \qquad (4.34)$$

From here, the particular solution can be seen only as a cosine wave of given amplitude A and phase η. Substituting eq. 4.34 into eq. 4.33, the following equation can be derived,

$$A = \frac{E_p}{\sqrt{(\omega_p^2 - \omega^2)^2 + c_p^2\omega^2}}, \quad \tan\eta = \frac{c_p\omega}{(\omega_p^2 - \omega^2)} \qquad (4.35)$$

A and η are important parameters for investigating **practical resonance**. In nature, damping and external forces exist such that resonance phenomenon introduced in the previous chapter may not occur in infinite time. Despite this, a peak in amplitude of the dependent variable may occur under various frequencies of external forcing and damping after a certain time. This physical manifestation which can be derived from the exact solutions of linear non-homogeneous ODEs may be referred to as practical resonance. Its investigation may be useful in design of various systems.

To illustrate the importance of A and η, eq. 4.35 which comes from the ODE, eq. 4.29, is investigated. Specifically, let us look at the effect of an external force (eq. 4.29 of a certain angular frequency ω to A/E_p [↪142] and η of eq. 4.35. For example, a system with a natural angular frequency of $\omega_p = 1$ [1/s] is given and is being investigated under five type of resistance c_p: $1/4, 1/2, 1.0, 2.0, 3.0$.

The following python script may be used to generate the required analyses. The first and second script loops and overlays plots for the five types of c_p to create Figure 4.4 and Figure 4.5.

↪**142** A/E_p represents how many times the sine amplitude of the particular solution is larger than that of the external sine wave.

```
1   import numpy as np
2   import matplotlib.pyplot as plt
3
4   wp = 1. #natural angular frequency
5   w = np.linspace(0.,2.,100)
6   lstyle=['solid','dotted','dashed',(0, (5,1)),'
        dashdot']
7   lstyle_counter=0
8   for cp in [0.25,0.5,1.,2.,3.]:
9           AoEp = 1/np.sqrt((wp**2.-w**2.)**2.+cp
            **2.*w**2.)
10          plt.plot(w,AoEp,color='k',
11                  linestyle=lstyle[lstyle_counter],
12                  label='$c_p$=%4.2f'%(cp))
13          lstyle_counter+=1
14  plt.xlim([0.,2.])
```

```
15   plt.ylabel(r'$A/E_p$')
16   plt.xlabel(r'$\omega$')
17   plt.legend()
18   plt.show()
```

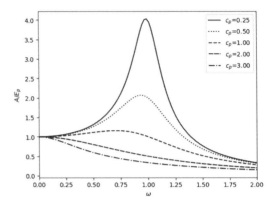

Figure 4.4   Sensitivity of the amplitude ratio to various set-ups of the RLC circuit

The python script for investigating the sensitivity of η:

```
1    import numpy as np
2    import matplotlib.pyplot as plt
3
4    wp = 1. #natural angular frequency
5    w = np.linspace(0.,2.,100)
6    lstyle=['solid','dotted','dashed',(0, (5,1)),'
        dashdot']
7    lstyle_counter=0
8    for cp in [0.25,0.5,1.,2.,3.]:
9            phase = np.arctan(cp*w/(wp**2.-w**2.))
10           phase[phase<0.] = phase[phase<0.]+np.pi
11           plt.plot(w,phase,color='k',
12                   linestyle=lstyle[lstyle_counter],
13                   label='$c_p$=%4.2f'%(cp))
14           lstyle_counter+=1
15   plt.xlim([0.,2.])
16   plt.ylabel(r'Phase')
17   plt.xlabel(r'$\omega$')
18   plt.legend()
```

First, let us consider the maximum value of amplitude ratio A/E_p from eq. 4.35 for a given external angular frequency ω. Differentiating eq. 4.35 with respect to ω,

$$\frac{\mathrm{d}(A/E_p)}{\mathrm{d}\omega} = -\frac{1}{2}((\omega_p^2 - \omega^2)^2 + \omega^2 c_p^2)^{-3/2} \times [2(\omega_p^2 - \omega^2)(-2\omega) + 2\omega c_p^2]$$

$$= 2\underbrace{((\omega_p^2 - \omega^2)^2 + \omega^2 c_p^2)^{-3/2}}_{1} \times \omega \times \underbrace{[(\omega_p^2 - c_p^2/2) - \omega^2]}_{2} \quad (4.36)$$

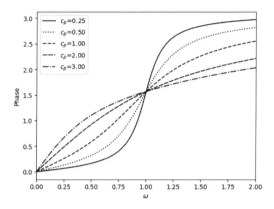

Figure 4.5 Sensitivity of the phase η to various set-ups of the RLC circuit

(1) will always be positive. Thus, (2) determines the concavity of A/E_p. If $\omega_p^2 - c_p^2/2 \leq 0$, (2) will always be negative, A/E_p will not have a positive peak value and will decrease under various ω (e.q. $c_p = 2.0, 3.0$). On the other hand if $\omega_p^2 - c_p^2/2 \geq 0$, an ω_{max} where A/E_p peaks can be determined by equating (2) to zero and isolating ω,

$$\omega_{max} = \sqrt{\omega_p^2 - c_p^2/2} \qquad (4.37)$$

From eq. 4.36 and eq. 4.37, it can be inferred that with decreasing resistance, the external angular frequency ω of maximum amplitude ratio approaches the value of the given natural angular frequency ω_p. In the case of $c_p = 0.25$, which is the minimum given resistance in the example, the maximum value of A/E_p (4) is situated close to the given $\omega_p (= 1)$.[143]

It is also clear from eq. 4.35 that regardless of the value of the given resistance, a phase shift of $\pi/2$ occurs when $\omega = \omega_p$.

↻143 Resonance will not occur under the presence of resistance even if $\omega_p = \omega$. Nevertheless, for small resistances under very close values of ω_p and ω, the amplitude ratio will remain significantly large. This state could damage the system (e.g. mass–spring, electric circuits). Therefore, a system in which the internal and external angular frequencies become equal will lead to a resonance-like phenomenon, which should be avoided.

4.4 Exercise 3

4.4.1 Investigating the behavior of a mass–spring system with external influence

Given 6 mass–spring systems of varying external force, solve for the ordinary differential equations. Analyze the behavior of the system by plotting the time series of the displacement of the point mass, the phase diagram of the point mass (Vertical axis) with its velocity (Horizontal axis). Note: the initial displacement and initial velocity is equal to 0 m and 0 m/s, respectively. The external force is $g(t) = F_0 \sin \omega t$ and the system has no resistance.

Exact solutions for the above system:

$\omega \neq \omega_p$:

Table 4.2 Mass–spring set-up

Set-up	Mass, m [kg]	Spring constant, k [kg/s^2]	Natural angular frequency $\omega_p = \sqrt{(k/m)}$ [1/s]	External force frequency ω[1/s]	External amplitude F_0[m/s^2]
(1)	1.0	π^2	π	0.25π	80
(2)	1.0	π^2	π	0.5π	80
(3)	1.0	π^2	π	2π	80
(4)	1.0	π^2	π	4π	80
(5)	1.0	π^2	π	1.1π	80
(6)	1.0	π^2	π	π	80

$$X = \frac{F_p}{\omega_p^2 - \omega^2}\left(-\frac{\omega}{\omega_p}\sin\omega_p t + \sin\omega t\right)$$

$$X' = \frac{F_p\omega}{\omega_p^2 - \omega^2}\left(-\cos\omega_p t + \cos\omega t\right)$$

$\omega = \omega_p$:

$$X = \frac{F_p}{2\omega_p^2}\left(\sin\omega_p t - t\omega_p\cos\omega_p t\right)$$

$$X' = \frac{F_p}{2}\left(t\sin\omega_p t\right)$$

4.4.2 Procedures for solving Exercise 3

In this exercise, python language will again be used. As with the previous exercise, the necessary modules are initially imported.

```
1  import matplotlib.pyplot as plt
2  import numpy as np
```

By carefully reading the problem of the exercise, the unchanging (constant) parameters can be initially defined. The parameters are mass m, spring constant k, and the amplitude of the external force F_0. Hence, the constant values corresponding to each parameters may be initially set.

```
1  #Constant parameters for all set-ups
2  m = 1.0
3  k = np.pi**2
4  Fo = 80.
```

Then, the time variable which is an array is set. For this exercise, a time interval of 0.01 s from 0 s to 50 s is set.

```
1  #time coordinates
2  t=np.arange(0,50.01,0.01)
```

Case 1 can now be set up. By looking at the relationship of ω and ω_p, an appropriate equation may be used (refer to the chapter or the supplementary notes of the exercise). The algorithm is constructed as follows.

```
1   #Case 1
2   #Parameters
3   wp = np.sqrt(k/m)
4   w = 0.25*np.pi
5   Fp = Fo/m
6   X = Fp*(-(w/wp)*np.sin(wp*t)+np.sin(w*t))/(wp*wp-w
    *w)
7   V = Fp*w*(-np.cos(wp*t)+np.cos(w*t))/(wp*wp-w*w)
```

Once the algorithm for the 1st case is set, a time-series is plotted.

```
1   plt.plot(t,X)
2   plt.xlabel('Time␣(s)')
3   plt.ylabel('Displacement␣(m)')
4   plt.title(r'$\omega=0.25\pi$␣and␣$\omega_p=\pi$')
5   plt.grid(True)
6   plt.show()
```

The time-series is generated as follows.

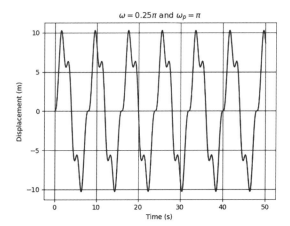

The cycle of the displacement is dominated by the angular frequency of the external force having a smaller value than the natural angular frequency (i.e. a longer cycle of the external force). From the figure, it can be seen that there is a small period ($=2\pi/\omega_p = 2$ s) corresponding to the cycle of the natural angular frequency.

Case 1's phase diagram is then plotted by using the following code (note: make sure to replace the previous code for the time-series with the one below to avoid overlapping figures).

```
1  plt.plot(V,X)
2  plt.xlabel('Velocity␣(m/s)')
3  plt.ylabel('Displacement␣(m)')
4  plt.title(r'$\omega=0.25\pi$␣and␣$\omega_p=\pi$')
5  plt.grid(True)
6  plt.show()
```

The phase diagram for case 1 is generated as follows.

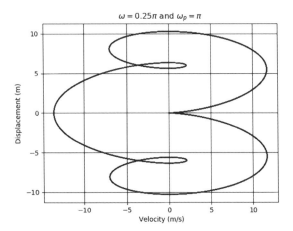

Unlike the case where no external forces are considered, the current system is no longer appearing as a simple circular curve orbiting around the origin. The loops are irregular. Since two periodic motions are superimposed, the displacement amplitude is smaller than the amplitude of the external force. This can also be understood by noticing the differences in velocities under the same displacement.

The same finding could be said for the 2nd case where the external angular frequency dominates the cycle. It can be briefly shown by the following code sequences. The resulting phase diagram also shows two values of velocity (of different magnitude) for the same displacement.

The time-series algorithm:

```
1  #Case 2
2  #Parameters
3  wp = np.sqrt(k/m)
4  w = 0.5*np.pi
5  Fp = Fo/m
6  X = Fp*(-(w/wp)*np.sin(wp*t)+np.sin(w*t))/(wp*wp-w
     *w)
7  V = Fp*w*(-np.cos(wp*t)+np.cos(w*t))/(wp*wp-w*w)
8  plt.plot(t,X)
9  plt.xlabel('Time␣(s)')
10 plt.ylabel('Displacement␣(m)')
11 plt.title(r'$\omega=0.5\pi$␣and␣$\omega_p=\pi$')
12 plt.grid(True)
13 plt.show()
```

The phase-diagram algorithm:

```
1   plt.plot(V,X)
2   plt.xlabel('Velocity (m/s)')
3   plt.ylabel('Displacement (m)')
4   plt.title(r'$\omega=0.5\pi$ and $\omega_p=\pi$')
5   plt.grid(True)
6   plt.show()
```

The time-series and phase diagram are plotted for case 2 as follows.

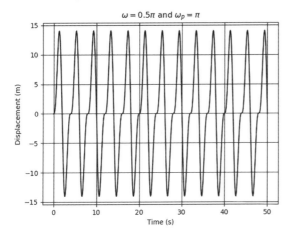

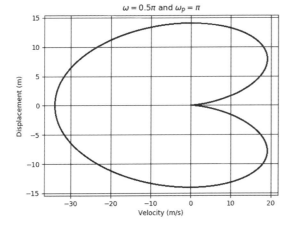

From case 3 and case 4, the conditions are reversed. The natural angular frequency is smaller than the external angular frequency ($\omega > \omega_p$). We can copy the code of case 1 and case 2 while replacing the value for the external angular frequency as follows.

```
1   #Case 3
2   #Parameters
3   wp = np.sqrt(k/m)
4   w = 2.*np.pi
5   Fp = Fo/m
6
```

```
 7   X = Fp*(-(w/wp)*np.sin(wp*t)+np.sin(w*t))/(wp*wp-w
     *w)
 8   V = Fp*w*(-np.cos(wp*t)+np.cos(w*t))/(wp*wp-w*w)
 9
10   plt.plot(t,X)  #Make sure to comment out this line
        when plotting the phase diagram.
11   plt.xlabel('Time␣(s)')
12   plt.ylabel('Displacement␣(m)')
13   plt.title(r'$\omega=2.\pi$␣and␣$\omega_p=\pi$')
14   plt.grid(True)
15   plt.show() #Make sure to comment out this line
        when plotting the phase diagram.
16
17   plt.plot(V,X)
18   plt.xlabel('Velocity␣(m/s)')
19   plt.ylabel('Displacement␣(m)')
20   plt.title(r'$\omega=2.\pi$␣and␣$\omega_p=\pi$')
21   plt.grid(True)
22   plt.show()
23
24   #Case 4
25   #Parameters
26   wp = np.sqrt(k/m)
27   w = 4.*np.pi
28   Fp = Fo/m
29
30   X = Fp*(-(w/wp)*np.sin(wp*t)+np.sin(w*t))/(wp*wp-w
     *w)
31   V = Fp*w*(-np.cos(wp*t)+np.cos(w*t))/(wp*wp-w*w)
32
33   plt.plot(t,X)  #Make sure to comment out this line
        when plotting the phase diagram.
34   plt.xlabel('Time␣(s)')
35   plt.ylabel('Displacement␣(m)')
36   plt.title(r'$\omega=4.\pi$␣and␣$\omega_p=\pi$')
37   plt.grid(True)
38   plt.show() #Make sure to comment out this line
        when plotting the phase diagram.
39
40   plt.plot(V,X)
41   plt.xlabel('Velocity␣(m/s)')
42   plt.ylabel('Displacement␣(m)')
43   plt.title(r'$\omega=4.\pi$␣and␣$\omega_p=\pi$')
44   plt.grid(True)
45   plt.show()
```

The time-series and phase diagram for Case 3 are as follows:

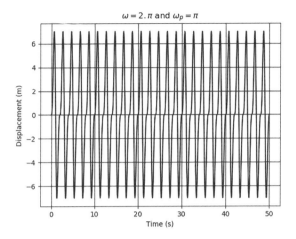

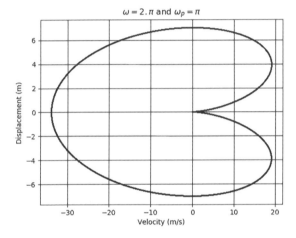

The time-series and phase diagram for Case 4 are as follows:

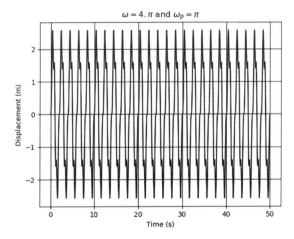

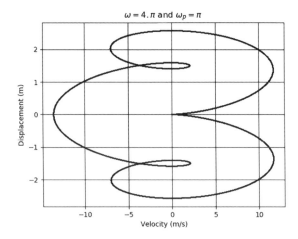

For cases 3 and 4, the natural angular frequency, which is smaller than the external force's angular frequency, now dominates the pattern of the time variation by shorter cycles compared to the 1st and 2nd case. The period of fluctuation is 2 s ($=\pi/\omega_p$). It can be seen that a smaller period corresponding to the period of the external force frequency exists in the behavior of the displacement. Looking at the phase diagram, an irregular loop trajectory is also seen similar to the previous cases. Since two periodic motions are superimposed (addition of two trigonometric functions), it can be understood that the velocity can have multiple values under the same displacement. Not only is the displacement amplitude smaller than the external force amplitude but also the amplitude of the displacement of the spring system decreases with increasing external force's angular frequency.

In case 5, the value of the external angular frequency is set to a value almost equal to the natural angular frequency. Setting the value ω or "w" to 1.1p, the following results are shown after running the following code.

```
1   #Case 5
2   #Parameters
3   wp = np.sqrt(k/m)
4   w = 1.1*np.pi
5   Fp = Fo/m
6
7   X = Fp*(-(w/wp)*np.sin(wp*t)+np.sin(w*t))/(wp*wp-w
       *w)
8   V = Fp*w*(-np.cos(wp*t)+np.cos(w*t))/(wp*wp-w*w)
9
10  plt.plot(t,X) #Make sure to comment out when
       plotting the phase diagram.
11  plt.xlabel('Time␣(s)')
12  plt.ylabel('Displacement␣(m)')
13  plt.title(r'$\omega=1.1\pi$␣and␣$\omega_p=\pi$')
14  plt.grid(True)
15  plt.show() #Make sure to comment out when plotting
       the phase diagram.
16
17  plt.plot(V,X)
18  plt.xlabel('Velocity␣(m/s)')
19  plt.ylabel('Displacement␣(m)')
20  plt.title(r'$\omega=1.1\pi$␣and␣$\omega_p=\pi$')
```

```
21  plt.grid(True)
22  plt.show()
```

The time-series and phase diagram are plotted for case 5 as follows.

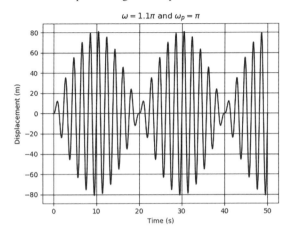

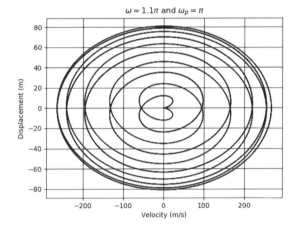

Recall the exact solution under the following assumptions $\varepsilon = \omega_p - \omega$, $\omega/\omega_p \cong 1$, $\omega + \omega_p \cong 2\omega_p$, $X = -(F_p/\varepsilon\omega_p)\sin(\varepsilon/2t)\cos\omega_p t$.

The pattern of the displacement in terms of its frequency and period is determined by the external angular frequency, the natural angular frequency, and its natural period which is approximately equal to 2 s $(= \pi/\omega_p)$. However, periodic fluctuation of the amplitude is observed at a period considerably longer the natural period. This demonstrates the "beats" phenomenon. The periodic fluctuation of the amplitude is commonly referred to as "swell." The angular frequency of the swell is likely to be $\varepsilon/2$ as shown from the equation $\sin(\varepsilon t/2)$, however it is twice that or equal to ε. That is, ε, or the difference between the external angular frequency and the natural angular frequency appears to determine the frequency of beats phenomena. Its period should be around 20 seconds. The reason behind this is that the sign of the first half of the cycle is opposite the sign of the latter half of the cycle. In this example, the maximum amplitude of the beats is about the same amplitude as the external force, which is about one order of magnitude larger than the systems (1) to (4). In the phase diagram, a plurality of loop trajectories can be observed.

Case 6 shows the behavior when $\omega = \omega_p$. Here, the second group of exact solution is used for X and V as follows.

```
1   #Case 6
2   wp = np.sqrt(k/m)
3   w = np.pi
4   Fp = Fo/m
5
6   X = Fp*(np.sin(wp*t)-t*wp*np.cos(wp*t))/2/wp/wp
7   V = Fp*(t*np.sin(wp*t))/2.
8
9   plt.plot(t,X)
10  plt.xlabel('Time␣(s)')
11  plt.ylabel('Displacement␣(m)')
12  plt.title(r'$\omega=\pi$␣and␣$\omega_p=\pi$')
13  plt.grid(True)
14  plt.show()
15  plt.plot(V,X)
16  plt.xlabel('Velocity␣(m/s)')
17  plt.ylabel('Displacement␣(m)')
18  plt.title(r'$\omega=\pi$␣and␣$\omega_p=\pi$')
19  plt.grid(True)
20  plt.show()
```

The time-series and phase diagram are plotted for case 5 as follows.

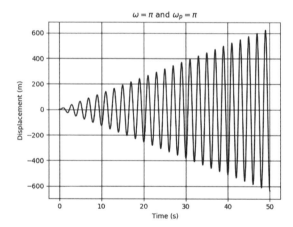

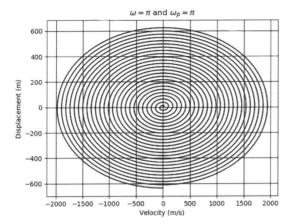

The behavior of the displacement is determined by either the external angular frequency and natural angular frequency. Its fluctuation cycle is about 2 s. However, its amplitude linearly increases with time which is consistent with the expected behavior of the second term of the exact solution. After a certain time, the displacement becomes significantly larger than the amplitude of the external force. In the phase diagram, a helical trajectory is shown with gradual increase in radius. This phenomenon is called "resonance." Try to imagine which physical phenomena manifest this behavior and based on the acquired knowledge from this chapter, try to imagine which specific parameters trigger this behavior.

❖❖❖❖❖❖ Problems for Exercise ❖❖❖❖❖❖

4.1   Find the exact solution of the 2nd order linear non-homogeneous ODE shown below. The internal parameters are assumed constant.

1. $T'' + 3T' + 2T = 8t^2$, $T(0) = T'(0) = 0$

2. $T'' - T' = 2t$, $T(0) = -1$, $T'(0) = 4$

3. $T'' + T = 4t$, $T(0) = -1$, $T'(0) = -1$

4. $T'' - 4T' + 3T = e^{2t}$, $T(0) = 0$, $T'(0) = 3$

5. $T'' - 4T' + 3T = e^t$, $T(0) = 0$, $T'(0) = 3$

6. $T'' - 4T' + 4T = e^{2t}$, $T(0) = 0$, $T'(0) = 3$

7. $T'' + 9T = \cos t$, $T(0) = 0$, $T'(0) = 3$

8. $T'' + 9T = 6\cos 3t$, $T(0) = 1$, $T'(0) = 3$

4.2   Consider the behavior of the mass–spring represented by the following ODE,

$$\frac{d^2 X}{dt^2} + \omega_p^2 X = F_p \sin \omega t.$$

1. Derive an equation based on the law of conservation of energy introduced in ch. 3. The work of the external force to the system is to be

expressed in integral form comprised of the function for external force and speed.

2. Substitute the exact solution to express the work done by the external force derived above. Confirm that energy is accumulated in the system.

4.3   In the mass–spring system with no resistance (introduced in ch. 4), find the general solution of the displacement X when the external force is replaced with $g(t) = F_0 e^{-2t}$. How will the system behave?

5 Numerics for ordinary differential equations

In this chapter, we will start to learn how to solve ODEs by numerics. Although numerics, in principle, can solve all ODEs, it has to be clearly stated that the methods are just approximations of the exact or analytics solutions. Unlike exact solutions of ODEs that are absolute, the solution obtained by numerics vary in accuracy depending on the settings of its implementation. The concrete numerical solution methods to be introduced are Euler's method and the Runge–Kutta method. The former is simple and yet inferior to the latter. In fact, the 4th order Runge–Kutta method is a solution widely used in practice. In each case, knowing the functional form of the first derivative (also referred to as the gradient of the tangent line) of a given ODE, the gradient is estimated for each interval of the independent variable to solve or predict the dependent variable. The Euler method uses only the slope of one tangent, but Runge–Kutta method advances time by weighted averaging slopes of four tangents at intervals close to the independent variable intervals.

5.1 Basic concept of the solution

Numerics is a practical tool for solving approximate solutions of all ODEs. There are several numerical solutions, but only two methods are introduced. The first method is the Euler method which is the most basic solution with poor accuracy. The second method is the Runge–Kutta method which is widely used. The steps are as follows ⌁**144**.

⌁**144** The fundamental concept is common for multiple numerical methods or approaches.

Step 1 Rearrange the ODE to be solved by isolating the first order as a function of the independent and dependent variables. The functional form of this first derivative means the slope of the tangent line in the time series diagram of the independent variable and the dependent variable. For example, given:

$$a\frac{dT}{dt} + bT + c = 0 \quad \text{becomes,}$$

$$g(t,T) = \frac{dT}{dt} = -\frac{b}{a}T - \frac{c}{a}.$$

The functional form of the slope of the tangent line is defined as $g(t,T)$.

Step 2 Unlike the exact solution, the numerical solution gives an approximate value of the dependent variable corresponding to the value of the specified independent variable. Therefore, the resolution (step size) of the independent variable is decided. The step size does not necessarily have to be a constant value. In this chapter, we assume the step size as constant[145].

Step 3   By using the Taylor's series expansion and the first derivative form in Step 1, an approximate solution to the ODE using basic arithmetics is established. This is called **discretization**. An approximate of the accuracy can be evaluated by the number of terms satisfied from the Taylor's series. From the initial value of the independent variable, the independent variable is incrementally increased while the dependent variable is solved by the derived basic arithmetic which comprises the gradient solved by substituting the previous independent variable and the corresponding dependent variable.

Step 4   In the case of higher order ODEs, it is necessary to first convert the ODEs to 1st order ODEs and reapply steps 1 to 3 to approximate the higher order terms[146].

5.2   Euler method

Let us solve an ODE comprised of temperature T (dependent variable) and time t (independent variable) as an example.

Step 1   Expression of the 1st derivative function (i.e. slope of tangent of time series)   If $f(T, \frac{dT}{dt}, t) = 0$ is the ODE, we modify this by expressing the first derivative as a function of $g(t, T)$ using the dependent variable T and the independent variable t.

$$\frac{dT}{dt} = g(t, T), \quad \text{Initial condition:} T(0) = T_0 \tag{5.1}$$

By assigning arbitrary independent variable t and dependent variable T to $g(t, T)$, the value of the first derivative is unambiguously determined. This value corresponds to the inclination of the tangent line in the time-series diagram of the independent variable (horizontal axis) and the dependent variable (vertical axis)[147].

Step 2   Specifying the resolution (step size)   Decide the step width Δt of time (independent variable) and starting from the initial state, sequentially obtain an approximate solution[148].

$$T(0) = T_0 \rightarrow T(\Delta t) \rightarrow T(2\Delta t) \rightarrow ... \rightarrow T(t) \rightarrow T(t + \Delta t) \rightarrow T(t + 2\Delta t) \rightarrow \cdots$$

For each step width, a serial number or index may be assigned to the corresponding independent variables and dependent variables such as the following,

$$t_0 = 0, t_1 = \Delta t, t_2 = 2\Delta t, ..., t_n = n\Delta t, ..., t_{n_{max}} = n_{max}\Delta t.$$
$$T_0 = T(0), T_1 = T(\Delta t), T_2 = T(2\Delta t), ...,$$

145   In practice, there are cases in which the required resolution or step size (e.g. time) is automatically determined while checking the calculation accuracy of the program.

146   Introduced in Sect. 5.5.

147   Recalling Figure 1.3, all slopes of the tangents of an independent variable for all particular solutions can be equal. Thus, even if the exact solution is not specified, the gradient (1st derivative of the dependent variable) of the solution can be specified.

148   Deciding the step size can be subjective. Small values will take more calculation time because more sequential calculations will be performed. Large values will reduce the accuracy of approximation.

$$T_n = T(n\Delta t),\ ...,\ T_{n_{\max}} = T(n_{\max}\Delta t) \qquad (5.2)$$

Step 3 Discretization   Based on Taylor's series, approximate a differential function that can be later estimated using basic arithmetic operators (addition, subtraction, division, and multiplication).

$$T(t+\Delta t) = T(t) + \frac{\Delta t}{1!}T'(t) + \frac{\Delta t^2}{2!}T''(t) + \frac{\Delta t^3}{3!}T'''(t) + \frac{\Delta t^4}{4!}T''''(t) + ...$$
$$(5.3)$$

Ignoring the higher order terms including and above the second order differential[149],

$$T(t+\Delta t) \cong T(t) + \Delta t\, T'(t). \qquad (5.4)$$

> **149** Only the first two terms provide a certain level of accuracy.

Substituting the first derivative of eq. 5.3 by $g(t,T)$ and the indexed variables into eq. 5.4, the independent variable can be estimated corresponding to the step size or interval of the dependent variable,

$$T_1 = T_0 + \Delta t\, g(t_0, T_0),$$
$$T_2 = T_1 + \Delta t\, g(t_1, T_1),$$
$$\vdots$$
$$T_{n+1} = T_n + \Delta t\, g(t_n, T_n). \qquad (5.5)$$

Starting from the initial value T_0, it is possible to obtain, $T_1, T_2, ..., T_n$, successively, by simple arithmetic without necessarily using iteration[150]. The numerical solution method introduced above is called the **Euler method**. Let us consider the meaning of the approximation from the plot of independent and dependent variables (Figure 5.1). Let the dependent variable T_n at the current time t_n be the starting time (t_n, T_n). In the time series chart, it is indicated by •. From that position, we advance in time using the step size Δt to determine the dependent variable T_{n+1} at the next time t_{n+1}. The value obtained by substituting the coordinate (t_n, T_n) of the starting point into the functional expression of the first derivative function defined in step 1 (i.e. $g(t_n, T_n)$), is the tangent of the curve (time-series plot) at that point. The tangent line can be extended from the starting point by Δt to determine T_{n+1}. As an analogy, imagine firing a gun at a certain angle defined by

> **150** In numerics, the differencing method which does not require iteration is called explicit method. On the other hand, implicit method is conducted by iteration while advancing the dependent variable.

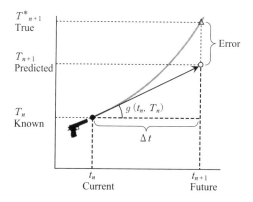

Figure 5.1   Conceptual diagram of Euler method

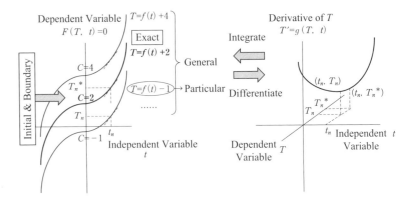

Figure 5.2   Graphical interpretation of numerical solution of first derivative value

$g(t_n, T_n)$ from the current position t_n, up to a distance Δt. The coordinates (t_{n+1}, T_{n+1}) of the predicted value are indicated by ∘ in the time-series chart. Even if the current dependent variable value at the starting point is a true value, the predicted value after Δt has an error relative to the true value of the next point (Δ in the figure). In practice, it should be noted that there is a possibility that errors gradually accumulate as the position departs sequentially from the initial position. These are brought about by ignoring the higher order terms of the Taylor series and the precision of calculation (rounding off numbers).

If the step size Δt takes a sufficiently small value, $\Delta t, \Delta t^2, \Delta t^3$, becomes much smaller. Therefore, the higher order derivative terms of the Taylor series will have a smaller contribution to the error. In the Euler method, the truncation error in each step is proportional to Δt^2, written $O(\Delta t^2)$, where O suggest order. Over a fixed interval in which the ODE is solved, the number of steps is proportional to $1/\Delta t$. The total error is proportional to $\Delta t^2(1/\Delta t) = \Delta t$. This is why Euler method is called a first-order method[151].

If an error is included in the approximate solution of the current position, its tangent $g(t_n, T_n)$ will also not be a tangent of the exact solution at the current position. Let us consider the meaning of $g(t_n, T_n)$[152].

Figure 5.2 compares the two-dimensional graph (left figure) of the dependent-independent variable with the 3-D graph (right figure) of the dependent variable (1st derivative, independent variable, dependent variable). The error of the approximate solution of the current time will not fall on the exact solution curve in the graph of dependent-independent variable but on a curve of another particular solution close to it. First derivatives $g(t_n, T_n)$ obtained by substituting arbitrary independent variable and dependent variable values are plotted on the 3-D graph on the right. For any value of (t_n, T_n), a unique value of slope or tangent line is acquired that can represent a particular solution[153]. The Euler method applied to a current variable value is close to an exact solution at that instant and belongs on a curve of a particular solution different from the exact solution. If the tangent line is used at that point, the future is estimated through extrapolation. This can be called a process of prediction. The next Runge–Kutta method uses four

151   The reason why accuracy is somehow achieved with only the first two terms is that the lower order terms have significant contribution to the approximated values.

152   By understanding its meaning, the principle behing the seemingly complicated Runge–Kutta method can be understood.

153   In the simple example shown in Figure 1.3, the slope of the tangent to the independent variable t at the position is an exact value since the 1st-order derivative of the independent variable T is given.

special solution tangents instead of 1 in order to increase its accuracy.

5.3   4th order Runge–Kutta

Like the Euler method, the 4th order ODE is used to numerically approximate the 1st order ODE where temperature T (dependent variable) depends on time t (independent variable). Step 1 and step 2 are exactly the same as the Euler method, but the differentiation method from step 3 differs because the higher order terms of the Taylor's series have to be included for better accuracy.

In the Euler method, the value of the dependent variable after Δt was predicted from only one slope determined from the current time. In the 4th order Runge–Kutta method, temporary prediction after Δt seconds is performed using 4 different slopes, and the prediction values after Δt seconds are determined by weighted averaging of predicted values[154]. As is evident from the error of the Euler method (Figure 5.1), it is preferable to use not only the current time information but also the information related to the future time such as the future time's gradient[155]. The problem is that the slope is also unknown because the dependent variable value of the future time is also unknown. To compromise, we can substitute the predicted value by the Euler method into the 1st derivative form determined from step 1, solve for the value of a new slope, and predict from the current time again using this new slope. Calculating various slopes from various prediction values are shown in the steps below and on Figure 5.3.

[154] In the Euler method, a slope of one particular solution was used. With unknown exact solution, it is intuitive to use more tangent information from other particular solutions to have better accuracy.

[155] If future values are required in the approximation, it is necessary to iterate calculations (implicit method) thereby increasing calculation time. It is advisable to conduct explicit method of numerics whenever possible.

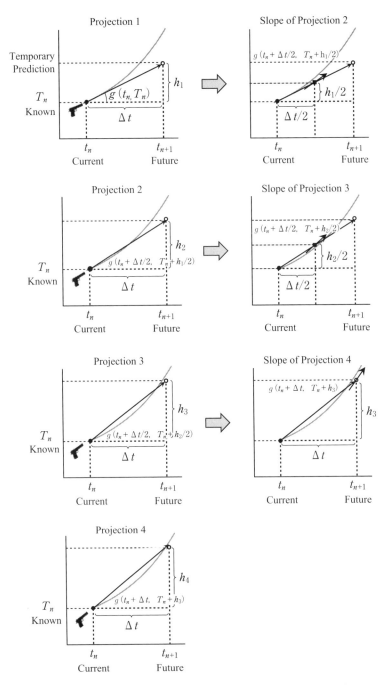

Figure 5.3 Conceptual diagram of Runge–Kutta method[156]

○**156** The coordinates (independent-dependent variable pairs) at the current time of projection 1 include errors if these coordinates do not correspond to the initial or boundary conditions. Thus, the curve on which the coordinates lie is not situated on the curve of the exact solution but rather close only to it. This is the same for projections 2 to 4. Although not plotted, each coordinates which arise from numerics lie on a particular solution curve.

Step 1 A temporary predicted value after Δt seconds is obtained by the Euler's method. The change amount after Δt seconds of the dependent variable from T_n is set as h_1.

$$h_1 = \Delta t\, g(t_n, T_n) \tag{5.6}$$

Step 2 Substitute the midpoint coordinates at $(t_n + \Delta t/2, T_n + h_1/2)$

which was taken from the current time coordinates and the predicted time coordinates after Δt seconds into the first derivative function and solve for the gradient $g(t_n + \Delta t/2, T_n + h_1/2)$. From the current time coordinates (t_n, T_n), the second trial prediction is performed using the new gradient to obtain the provisional predicted value after Δt seconds. Let the change after Δt seconds of the dependent variable T_n be denoted by h_2[157].

$$h_2 = \Delta t\, g(t_n + \Delta t/2, T_n + h_1/2) \tag{5.7}$$

Step 3   Substitute the midpoint coordinates at $(t_n + \Delta t/2, T_n + h_2/2)$ which was taken from the current time coordinates and the predicted time coordinates after Δt seconds into the first derivative function and solve for the gradient $g(t_n + \Delta t/2, T_n + h_2/2)$. From the current time coordinates (t_n, T_n), the third trial prediction is performed using the new gradient to obtain the provisional predicted value after Δt seconds. Let the change after Δt seconds of the dependent variable T_n be denoted by h_3.

$$h_3 = \Delta t\, g(t_n + \Delta t/2, T_n + h_2/2) \tag{5.8}$$

Step 4   Substitute the coordinates at $(t_n + \Delta t, T_n + h_3)$ which was taken from the current time coordinates and the predicted time coordinates after Δt seconds into the first derivative function and solve for the gradient $g(t_n + \Delta t, T_n + h_3)$. From the current time coordinates (t_n, T_n), the fourth trial prediction is performed using the new gradient to obtain the provisional predicted value after Δt seconds. Let the change after Δt seconds of the dependent variable T_n be denoted by h_4[158].

$$h_4 = \Delta t\, g(t_n + \Delta t, T_n + h_3) \tag{5.9}$$

Step 5   The four provisional predictions for T_{n+1} are weighted and averaged as follows in order to determine the value of the dependent variable after Δt seconds.

$$T_{n+1} = T_n + \frac{1}{6}(h_1 + 2h_2 + 2h_3 + h_4) \tag{5.10}$$

Actually, the 4th order Runge–Kutta method satisfied the terms of the Taylors Series up to the Δt^4 term thus giving it an accuracy up to the 4th order[159]. The coefficients in the weighting and the coordinates used for the provisional values for the dependent variable were decided to ensure 4th order accuracy of the method. Incidentally, other combinations that lead to a 4th order accuracy are also available.

In the numerics of 1st order ODE, there is no restriction on the functional form $g(t, T)$[160]. Whether $g(t, T)$ is non-linear or non-homogeneous, the procedure to approximating the solution is the same. However, as the value of the independent variable being analyzed departs form the known initial condition, errors accumulate rapidly, resulting in large deviations of the predicted value from the true value (divergence of numerical solutions). In addition, there is also a possibility that a solution with an error can not be

157   From the right-side figure of Figure 5.2, variable values that differ from the current time are substituted to the slope equation. This leads to a slope of a particular solution curve that is different from the Euler method (left-side figure).

158   The slopes refer to four different particular solution curves (Figure 5.2) at the following independent and dependent variable coordinates:
(t_n, T_n)
$(t_n + \Delta t/2, T_n + h_1/2)$
$(t_n + \Delta t/2, T_n + h_2/2)$
$(t_n + \Delta t, T_n + h_3)$

159   By applying Taylor-series expansion to eq. 5.6 to 5.10 and comparing them with the 4th order terms in eq. 5.3, the 4th order accuracy is obtained. See the problem exercise of this chapter.

160   If the functional form of the ODE is known, the value of the derivative is calculated by substituting known pairs of independent variables and dependent variables.

practically accepted even if the above steps are followed. Caution must be done when using numerics.

5.4   Numerical solution of multiple simultaneous 1st order ODEs

▷161   To show that numerics are applicable to any ODE, this section applies it to a case where multiple dependent variables are involved.

Let us consider an example of a binary simultaneous 1st order ODE[▷161]. Assume that time (t) is an independent variable, while temperature (T) and CO_2 (C) related dependent variables which are represented by 2 ODEs.

Step 1   Express in terms of a 1st order ODE

$$\frac{dT}{dt} = g_1(t, T, C), \quad \text{Initial condition } T(0) = T_0.$$

$$\frac{dC}{dt} = g_2(t, T, C), \quad \text{Initial condition } C(0) = C_0. \quad (5.11)$$

The difference with the case of only one dependent variable is that the 1st

▷162   This can be imagined by replacing the 3-D graph in Figure 5.2 with a 4-D graph of 1-independent variable and 2-dependent variables. If the coordinate of a dependent variable is given, the value of the 1st derivative for both dependent variables can be calculated as in the previous section.

order derivative function contains another dependent variable[▷162].

Step 2   Setting the resolution   Independent and dependent variables are identified by adding a serial number or index to each.

$$t_0 = 0, t_1 = \Delta t, t_2 = 2\Delta t, ..., t_n = n\Delta t, ..., t_{n_{max}} = n_{max}\Delta t.$$

$$T_0 = T(0), T_1 = T(\Delta t), T_2 = T(2\Delta t), ..., T_n = T(n\Delta t), ..., T_{n_{max}} = T(n_{max}\Delta t).$$

$$C_0 = C(0), C_1 = C(\Delta t), C_2 = C(2\Delta t), ..., C_n = C(n\Delta t), ...C_{n_{max}} = C(n_{max}\Delta t).$$

$$(5.12)$$

▷163   Euler method can be used following Sect. 5.2.

Step 3   Differencing by Runge–Kutta method[▷163]   Although differencing will be done for 2 dependent variables, the steps for numerical analyses of a single variable applies.

Projection 1   A temporary predicted value after Δt seconds is obtained by the Euler method. The amount of change after Δt seconds of C_n and T_n, are denoted by h_1 and k_1, respectively.

$$h_1 = \Delta t g_1(t_n, T_n, C_n),$$

$$k_1 = \Delta t g_2(t_n, T_n, C_n). \quad (5.13)$$

Projection 2   Using the Euler method, predict the value of increase at the midpoint positioned between the current time and the predicted time $(t_n + \Delta t/2, T_n + h_1/2, C_n + k_1/2)$ and calculate the slopes at that point. The second trial is performed using the coordinates (t_n, T_n, C_m) as the starting point and the earlier derived slope to determine the change after Δt seconds. The change after Δt seconds of T_n and C_n are referred to as h_2 and k_2, respectively.

$$h_2 = \Delta t g_1(t_n + \frac{\Delta t}{2}, T_n + \frac{h_1}{2}, C_n + \frac{k_1}{2}),$$

$$k_2 = \Delta t g_2(t_n + \frac{\Delta t}{2}, T_n + \frac{h_1}{2}, C_n + \frac{k_1}{2}). \quad (5.14)$$

Projection 3   Using the Euler method and the result from the previous projection, predict the value of increase at the midpoint positioned between the current time and the predicted time $(t_n + \Delta t/2, T_n + h_2/2, C_n + k_2/2)$ and calculate the slopes at that point using the first derivative function. From the current coordinates (t_n, T_n, C_n), the third trial is performed using the updated slope obtained earlier to determine the change after Δt seconds. The change after Δt seconds of T_n and C_n are referred to as $h3$ and $k3$ respectively.

$$h_3 = \Delta t g_1(t_n + \frac{\Delta t}{2}, T_n + \frac{h_2}{2}, C_n + \frac{k_2}{2}),$$

$$k_3 = \Delta t g_2(t_n + \frac{\Delta t}{2}, T_n + \frac{h_2}{2}, C_n + \frac{k_2}{2}). \tag{5.15}$$

Projection 4   Using the Euler method and the result from the previous projection 3, predict the value of increase at $(t_n + \Delta t, T_n + h_3, C_n + k_3)$ after substituting the results from projection 3 into the first derivative function. Using the coordinates of the current time (t_n, T_n, C_n) as the starting time, the fourth trial is performed using the derived slope to find the provisional value after Δt seconds. Let the amount of change after Δt seconds from T_n and C_n be denoted by h_4 and k_4, respectively.

$$h_4 = \Delta t g_1(t_n + \Delta t, T_n + h_3, C_n + k_3),$$

$$k_4 = \Delta t g_2(t_n + \Delta t, T_n + h_3, C_n + k_3). \tag{5.16}$$

Determination of predicted value   After acquiring four different provisional predictions, their contribution to the final prediction after Δt seconds is weighted as follows,

$$T_{n+1} = T_n + \frac{1}{6}(h_1 + 2h_2 + 2h_3 + h_4),$$

$$C_{n+1} = C_n + \frac{1}{6}(k_1 + 2k_2 + 2k_3 + k_4). \tag{5.17}$$

This method is applicable even for multiple simultaneous ODEs with more than 2 dependent variables[164].

5.5   Numerical solution of *n*-order ODEs

In the case of *n*-order ODEs, the solution method introduced in the previous section is applicable by replacing the *n*th order with 1st order ODEs. It is demonstrated by the following.

5.5.1   2nd order ODEs

The exact solution of the following equation cannot be easily obtained[165].

$$\frac{d^2 T}{dt^2} + a(t)\left[\frac{dT}{dt}\right]^n T^m + b(t)\left[\frac{dT}{dt}\right]^l + c(t)T^p = d(t), \tag{5.18}$$

Initial condition: $T(0) = T_0, T'(0) = T_0'$.

Here, the first order derivative of the dependent variable T is regarded as

▷**164**   In case of 3 dependent variables, an additional variable Y_n appears as follows,
$h_1 = \Delta t g_1(t_n, T_n, C_n, Y_n)$
$k_1 = \Delta t g_2(t_n, T_n, C_n, Y_n)$
$l_1 = \Delta t g_3(t_n, T_n, C_n, Y_n)$
....
$T_{n+1} = T_n + \frac{1}{6}(h_1 + 2h_2 + 2h_3 + h_4)$
$C_{n+1} = C_n + \frac{1}{6}(k_1 + 2k_2 + 2k_3 + k_4)$
$Y_{n+1} = Y_n + \frac{1}{6}(l_1 + 2l_2 + 2l_3 + l_4)$

▷**165**   The advantage of numerics is that even for complicated ODEs with non-linear coefficients, the algorithms do not need to be modified.

a new dependent variable C as shown,

$$\frac{dT}{dt} = C, \quad \text{Initial condition: } T'(0) = T_0' = C(0) = C_0. \tag{5.19}$$

Substituting eq. 5.19 into eq. 5.18, simultaneous ODEs will appear. That is,

$$\frac{dT}{dt} = g_1(t,T,C) = C, \quad \text{Initial condition: } T(0) = T_0.$$
$$\frac{dC}{dt} = g_2(t,T,C) = -a(t)C^n T^m - b(t)C^l - c(t)T^p + d(t),$$
$$\text{Initial condition: } C(0) = C_0. \tag{5.20}$$

The applicability of the method introduced in the previous section can be clearly seen when comparing the above equations with eq. 5.11. The 1st order derivatives of two dependent variables T and C are represented by functions $g_1(t,T,C)$ and $g_2(t,T,C)$.

5.5.2   3rd order ODE

To demonstrate the method further, let us assume an even higher order ODE such as the following of 3rd order,

$$\frac{d^3 T}{dt^3} + a(t)\frac{d^2 T}{dt^2} + b(t)\frac{dT}{dt} + c(t)T = d(t), \tag{5.21}$$
$$\text{Initial condition: } T(0) = T_0, T'(0) = T_0', T''(0) = T_0''.$$

The 1st and 2nd derivative of the dependent variable T can be regarded as new dependent variables C and V, respectively, as shown below.

$$\frac{dT}{dt} = C, \quad \text{Initial condition: } T'(0) = T_0' = C(0) = C_0,$$

$$\frac{d^2 T}{dt^2} = \frac{dC}{dt} = V, \quad \text{Initial condition: } T''(0) = T_0'' = V(0) = V_0. \tag{5.22}$$

Substituting eq. 5.22 into eq. 5.21 replaces the higher-order ODE with three-way simultaneous ODE. That is,

$$\frac{dT}{dt} = g_1(t,T,C,V) = C, \quad \text{Initial condition: } T(0) = T_0.$$
$$\frac{dC}{dt} = g_2(t,T,C,V) = V, \quad \text{Initial condition: } C(0) = C_0.$$
$$\frac{dV}{dt} = g_3(t,T,C,V) = -a(t)V - b(t)C - c(t)T + d(t),$$

$$\text{Initial condition: } V(0) = V_0. \tag{5.23}$$

▷**166** Of the N number of 1st-order ODEs, $(N-1)$ terms become new dependent variables (above two equations of eq. 5.23) and only one differential equation remains which is equated to the functional form of the Nth order ODE.

As clearly shown above, higher order ODEs can be approximately solved using numerics by simply replacing its derivatives (1st to $n-1$) with other new dependent variables to form a simultaneous ODE▷**166**.

5.6   Exercise 4

5.6.1   Applying the Euler and Runge–Kutta to the logistic model

From the 1st order non-linear non-homogeneous ODE (logistic curve) shown below,

$$\frac{dN}{dt} = (a_3 - a_4 N)N$$

where N is the number of organisms, a_3 is the growth rate (0.05/day), and a_4 is the carrying capacity (0.00005/organism/day), solve for the population using the exact solution, the Euler method, Heun method (Problems for Exercise 5.1), and 4th order Runge–Kutta method. Plot the time-series of population. The units of t is in days. The initial number of organisms is 100.

Additional instructions: Construct a time-series plot up to 120 days with 1-day time interval.

Recall the exact solution:

$$N = \frac{a_3}{a_4 + C\exp(-a_3 t)}, \quad C = \frac{a_3 - a_4 N_0}{N_0}$$

5.6.2   Applying the Euler and Runge–Kutta to the mass–spring system

Apply the Euler and 4th order Runge–Kutta method to the mass–spring system governed by the following 2nd order ODE,

$$\frac{d^2 X}{dt^2} + \omega_p^2 X = F_p \sin \omega t, \quad X(0) = X'(0) = 0.$$

Table 5.1   Mass–spring set-up

Natural angular frequency $\omega_p[1/s]$	External angular frequency $\omega[1/s]$	External amplitude $F_p[m/s^2]$
π	2π	80

1.   Plot a time-series of displacement X (m) from time t=0,2s. Compare the calculated displacement using the exact solution, Euler method, and 4th order Runge–Kutta method. In addition, see the accuracy of the numerical method when the time interval is changed from 0.01s, 0.05s, and 0.1s.

2.   Plot a time-series and phase diagram of displacement X (m) from time t=0,10s. Compare the result when the displacement is calculated using the exact solution, Euler method, and 4th order Runge–Kutta method. Use a time interval of 0.01 s.

Recall the exact solution when $\omega_p \neq \omega$ and $X(0) = X'(0) = 0$ is

$$X = \frac{F_p}{\omega_p^2 - \omega^2} \left(-\frac{\omega}{\omega_p} \sin \omega_p t + \sin \omega t \right)$$

$$X' = \frac{F_p \omega}{\omega_p^2 - \omega^2} \left(-\cos \omega_p t + \cos \omega t \right)$$

5.6.3 Procedures for solving Exercise 4

◆ **Procedure for Sect. 5.6.1**

Here are the steps to constructing the program. First, construct the necessary functions for the numeric algorithms and the derivatives. Begin by importing the modules.

```
1  import numpy as np
2  import matplotlib.pyplot as plt
```

Next, define the functions for the exact, Euler, Heun, and 4th order Runge–Kutta methods. Notice that the function for the slope is also constructed (i.e. not individually inside the numerics).

```
1  def exact(t,No):
2          a3 = 0.05
3          a4 = 0.00005
4          c = (a3-a4*No)/No
5          N = a3/(a4+c*np.exp(-a3*t))
6          return N
7  def euler(t,No,delta):
8          N = np.copy(t)
9          N[0] = No
10         for i in range(0,np.shape(t)[0]-1):
11                 N[i+1] = N[i]+delta*dndt(N[i])
12         return N
13 def heun(t,No,delta):
14         N = np.copy(t)
15         N[0] = No
16         for i in range(0,np.shape(t)[0]-1):
17                 h1 = delta*dndt(N[i])
18                 h2 = delta*dndt(N[i]+h1)
19                 N[i+1] = N[i]+0.5*(h1+h2)
20         return N
21 def rk(t,No,delta):
22         N = np.copy(t)
23         N[0] = No
24         for i in range(0,np.shape(t)[0]-1):
25                 h1 = delta*dndt(N[i])
26                 h2 = delta*dndt(N[i]+h1/2.)
27                 h3 = delta*dndt(N[i]+h2/2.)
28                 h4 = delta*dndt(N[i]+h3)
29                 N[i+1] = N[i]+(h1+2.*h2+2.*h3+h4)
                        /6.
30         return N
31 def dndt(N):
32         a3 = 0.05
33         a4 = 0.00005
34         return (a3-a4*N)*N
```

Next, construct the main program as follows. Instead of solving the values and plotting, the calculations can be done inside the plot object. See the concise program below. Carefully deduce the definition of the variables used (e.g. delta is the step size).

```
1   delta = 1.
2   t = np.arange(0.,120.+delta,delta)
3   No = 100
4   plt.plot(t,exact(t,No),'r',label='Exact')
5   plt.plot(t,euler(t,No,delta),'g',label='Euler')
6   plt.plot(t,heun(t,No,delta),'k',label='Heun')
7   plt.plot(t,rk(t,No,delta),'y',label='4th␣Runge-
    Kutta')
8   plt.legend(loc=0)
9   plt.xlabel('Days')
10  plt.ylabel('Organism␣Population')
11  plt.show()
```

Running the program, will result in the following figure.

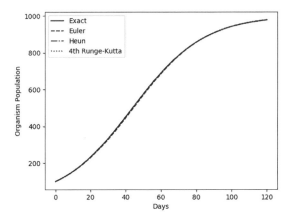

Notice that all numerical models approximately fit the analytical solution! If an error is encountered, it is most likely that a variable was misspelled, the closing parentheses are lacking, or the code was not properly indented. The differences of each numerical method will be more apparent in the next exercise.

◆ Procedure for Sect. 5.6.2

Again, begin by importing the necessary modules.

```
1   import numpy as np
2   import matplotlib.pyplot as plt
```

The next step is to define the functions as before. Modifications to the earlier steps are needed since in this second exercise, a 2nd order ODE is being dealt. To model this numerically, the 2nd order ODE ($d^2X/dt^2 + \omega_p^2 X = F_p \sin \omega t$) can be rewritten into two 1st order ODEs. The first is obtained by setting $d^2X/dt^2 = dV/dt$ which will lead to $dX/dt = V \, dV/dt$. The second

1st order ODE, $\mathrm{d}V/\mathrm{d}t = -\omega_p^2 + F_p \sin \omega t$ can be derived by substituting the earlier defined 2nd order ODE. The numerical algorithm is then constructed based on the solution to the two simultaneous 1st order ODEs. Separate functions for the two derivatives are thus written as follows.

```
1   def exact(t):
2           wp = np.pi
3           w = 2.*np.pi
4           Fp = 80.
5           X = Fp*((-w*np.sin(wp*t)/wp)+np.sin(w*t))
                /(wp*wp-w*w)
6           V = Fp*w*(-np.cos(wp*t)+np.cos(w*t))/(wp*
                wp-w*w)
7           return X,V
8   def euler(t,delta):
9           X = np.copy(t)
10          V = np.copy(t)
11          X[0] = 0.
12          V[0] = 0.
13          for i in range(0,np.shape(t)[0]-1):
14                  X[i+1] = X[i]+delta*dxdt(V[i])
15                  V[i+1] = V[i]+delta*dvdt(t[i],X[i
                        ])
16          return X,V
17  def heun(t,delta):
18          X = np.copy(t)
19          V = np.copy(t)
20          X[0] = 0.
21          V[0] = 0.
22          for i in range(0,np.shape(t)[0]-1):
23                  h1 = delta*dxdt(V[i])
24                  k1 = delta*dvdt(t[i],X[i])
25                  h2 = delta*dxdt(V[i]+k1)
26                  k2 = delta*dvdt(t[i]+delta,X[i]+h1
                        )
27                  X[i+1] = X[i]+0.5*(h1+h2)
28                  V[i+1] = V[i]+0.5*(k1+k2)
29          return X,V
30  def rk(t,delta):
31          X = np.copy(t)
32          V = np.copy(t)
33          X[0] = 0.
34          V[0] = 0.
35          for i in range(0,np.shape(t)[0]-1):
36                  h1 = delta*dxdt(V[i])
37                  k1 = delta*dvdt(t[i],X[i])
38                  h2 = delta*dxdt(V[i]+k1/2.)
39                  k2 = delta*dvdt(t[i]+delta/2.,X[i
                        ]+h1/2.)
40                  h3 = delta*dxdt(V[i]+k2/2.)
41                  k3 = delta*dvdt(t[i]+delta/2.,X[i
                        ]+h2/2.)
42                  h4 = delta*dxdt(V[i]+k3/2.)
43                  k4 = delta*dvdt(t[i]+delta,X[i]+h3
                        /2.)
44                  X[i+1] = X[i]+(h1+2.*h2+2.*h3+h4)
                        /6.
45                  V[i+1] = V[i]+(k1+2.*k2+2.*k3+k4)
                        /6.
46          return X,V
47  def dxdt(V):
```

```
48              return V
49  def dvdt(t,X):
50              wp = np.pi
51              w = 2.*np.pi
52              Fp = 80.
53              return -wp*wp*X+Fp*np.sin(w*t)
```

The above functions can then be used to complete the required tasks for this exercise. The first requirement is then completed using the following code. In the code, looping is introduced to repeat each procedure for each case smoothly.

```
1   #Requirement 1
2   plt.figure(1, figsize=(12,6))
3   plt.subplot(1,2,1)
4   delta = 0.01
5   t=np.arange(0.,2.+delta,delta)
6   X,V=exact(t)
7   plt.plot(t,X,'r',label='exact')
8   plt.xlabel('days')
9   plt.ylabel('displacement')
10  plt.title('Euler')
11  for delta in [0.01,0.05,0.1]:
12              t=np.arange(0.,2.+delta,delta)
13              X,V=euler(t,delta)
14              plt.plot(t,X,label=delta)
15  plt.legend(loc=0)
16  plt.subplot(1,2,2)
17  delta = 0.01
18  t=np.arange(0.,2.+delta,delta)
19  X,V=exact(t)
20  plt.plot(t,X,'r',label='exact')
21  plt.xlabel('days')
22  plt.ylabel('displacement')
23  plt.title('Runge-Kutta')
24  for delta in [0.01,0.05,0.1]:
25              t=np.arange(0.,2.+delta,delta)
26              X,V=rk(t,delta)
27              plt.plot(t,X,label=delta)
28              plt.subplot(1,2,2)
29  plt.legend(loc=0)
30  plt.show()
```

Running the program, will result in the following figure.

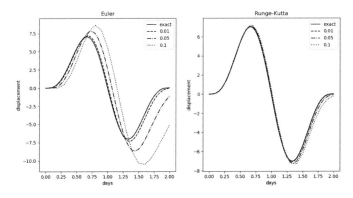

By analyzing the time-series plot, it can be seen how inaccurate the Euler method is compared to the 4th order Runge–Kutta. As time (days) increases, the larger step size or time resolution cases depart from the exact solution! Interestingly, there is little change in accuracy for the 4th order Runge–Kutta in spite of the large differences in step size. It can also be deduced from the algorithm that refining the step size from 0.1 to 0.01 also increases the calculation time since the number of loops is increased inside each function.

The same looping technique can be done to accomplish the second task. Instead of equating the function (since it returns two variables from the "def" code) to two variables (e.g. X,V =euler(t,delta)), euler(t,delta)[0] is equivalent to X and euler(t,delta)[1] is equivalent to V.

```
1   delta = 0.01
2   t=np.arange(0.,10.+delta,delta)
3   plt.figure(2, figsize=(6,12))
4   cases = [(t,exact(t)[0],'Exact'),(exact(t)[1],
      exact(t)[0],'Exact')
5   ,(t,euler(t,delta)[0],'Euler'),(euler(t,delta)[1],
      euler(t,delta)[0],'Euler')
6   ,(t,rk(t,delta)[0],'4th␣Runge-Kutta'),(rk(t,delta)
      [1],rk(t,delta)[0],
7   '4th␣Runge-Kutta')]
8   for index,(X,Y,tt) in enumerate(cases):
9           plt.subplot(3,2,index+1)
10          plt.plot(X,Y,'k-')
11          plt.title(tt)
12  plt.show()
```

Running the program, will result in the following figure.

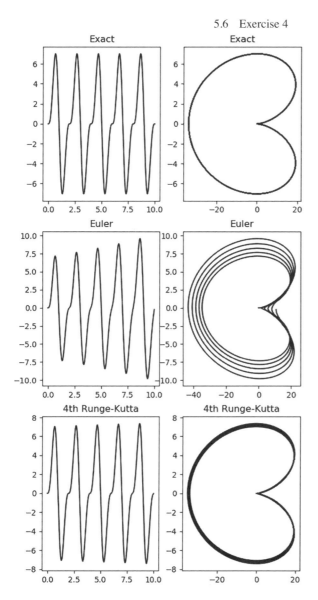

As shown from the result, it can be seen that refining the resolution to 0.01 will lead to comparable results between the Euler and 4th order Runge–Kutta method. It is obvious that the errors gradually accumulate with increasing time (independent variable) such as shown by the increasing maximum amplitude of displacement and velocity. Although the behavior is not obvious in the time-series diagram, it can be seen from the phase diagram that the constant two-period behavior was not satisfied since the trajectory increases with every cycle (expanding shape). Feel free to modify the code (e.g. introduce Heun method by referring to Problems for Exercise 5.1).

◆◆◆◆◆◆ **Problems for Exercise** ◆◆◆◆◆◆

5.1   Proving whether the 4th order Runge–Kutta is indeed of 4th-order accuracy using the Taylor-series expansion is complicated. Here, let us settle for the 2nd order Runge–Kutta method.

The provision prediction values from two trials are defined as follows,

$$h_1 = \Delta t \times g(t_n, T_n) \tag{5.24a}$$

$$h_2 = \Delta t \times g(t_n + \Delta t, T_n + h_1) \tag{5.24b}$$

In order to satisfy a 2nd-order accuracy of the Taylor-series expansion using the following equation, determine the value of the weighing factors, α and β.

$$T_{n+1} = T_n + (\alpha h_1 + \beta h_2) \tag{5.25}$$

5.2   Find the exact and numerical solutions of the following ODE and compare the accuracy of the solutions. For the numerical solutions, use Python language with a time step of 0.1 s.

$T' + 5T = 0, \quad T(0) = 2$

6 Ordinary differential equations and chaos phenomena

6.1 Non-linear mass–spring system and chaos

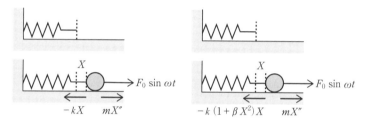

Figure 6.1 Linear springs (left) and non-linear springs (right) acted upon by an external force. Figures on the top are natural springs. Friction resistance is assumed negligible at the area of contact between the mass and the floor.

From the mass–spring system experiencing an external force, let us analyze a more complex restorative phenomenon resulting from a common characteristic of springs to not necessarily follow Hooke's law (i.e. k is replaced by a non-linear parameter). Consider a side-ways mass–spring system as shown in Figure 6.1[167] where gravity has no influence in the spring's motion throughout. The difference with the previously discussed mass–spring system is that the string is not elongated during its stationary equilibrium state. In this example, the resistance generated at the area of contact between the mass and floor is ignored. Assuming displacement X as the dependent variable, the behavior of the mass–spring system can be expressed by an ODE (eq. 6.1). The second term on the left side represents the restoring force according to Hooke's law.

$$m\frac{\mathrm{d}^2 X}{\mathrm{d}t^2} + kX = F_0 \sin \omega t \qquad (6.1)$$

(Internal system)   (External system)

Next, a non-linear spring that does not follow Hooke's law is considered. The restoring force of the spring represented by the product of a spring coefficient and the displacement is common to both linear and non-linear

[167] The ODE of a non-linear mass–spring system under the same configuration as that of ch. 4 will have a gravity term because the elongation of the spring in stationary state will not balance with the gravity term. For simplicity of interpretation, let us assume having no effect to gravity by slightly modifying the non-linear mass–spring system.

springs. In the case of a linear spring, the spring coefficient is constant; in the case of non-linear spring, the spring coefficient is a function of the displacement. Let us consider a virtual non-linear spring where the spring coefficient is a function of the square of the displacement (eq. 6.2)[168]. The internal parameter β expresses the strength of non-linearity of the spring coefficient. If β is 0, the spring is essentially a linear spring. As the value β increases, the stronger the restoring force of the spring proportional to the square of the displacement.

□168 Such a spring does not exist but is assumed here to provide a clear picture of non-linear behavior. The spring constant is not necessarily proportional to the square of the displacement. Note as well that non-linearity will appear regardless of whether the spring constant is proportional to the 1st power or 3rd power of the displacement.

$$\text{Linear spring coefficient} \to k$$
$$\text{Non-linear spring coefficient} \to k(1+\beta X^2) \text{[169]}$$

□169 Here, β is an internal parameter which determines the strength of non-linearity. If β is 0, the system will behave linearly. The higher the value of β, the more non-linear the system becomes.

Introducing the non-linear spring coefficient, results in a non-linear ODE as follows,

$$m\frac{\mathrm{d}^2 X}{\mathrm{d}t^2} + k(1+\beta X^2)X = F_0 \sin \omega t \qquad (6.2)$$

$$\text{(Internal system)} \quad \text{(External system)}$$

Substituting $\omega_p = \sqrt{k/m}$ and $F_p = F_0/m$ into the equation,

$$\frac{\mathrm{d}^2 X}{\mathrm{d}t^2} + \omega_p^2(1+\beta X^2)X = F_p \sin \omega t \qquad (6.3)$$

Since this expression includes a non-linear term at the second term of the left-side of the equation, obtaining an exact solution is almost impossible and tedious. For this equation, numerics can be used. In order to apply the Runge–Kutta method, the ODE (eq. 6.3) must be converted to a first order simultaneous ODE as follows[170],

□170 Recall the previous chapter on numerical approximation of higher-order ODEs.

$$\begin{aligned} \frac{\mathrm{d}X}{\mathrm{d}t} &= V, \\ \frac{\mathrm{d}V}{\mathrm{d}t} &= -\omega_p^2 X - \beta \omega_p^2 X^3 + F_p \sin \omega t. \end{aligned} \qquad (6.4)$$

An interesting behavior of the displacement is manifested in the non-linear system. This behavior is called "chaos." According to the book "A Survey of Computational Physics: Introductory Computational Science" by Bordeianu and Landau, chaos is a deterministic behavior of a system displaying no discernible regularity. Before investigating the behavior of the above mass–spring system, it is necessary to define chaos and its characteristics.

6.2   Physical definition and characteristics of chaos

Unlike linear ODEs, non-linear ODEs tend to result in multi-phase solutions or chaos. The difficulty of non-linear ODEs is not only because its exact solution is extremely difficult to derive but because of the diversity and complexity of the physical phenomena it entails. Here, the general features and physical meaning of chaos are summarized.

(1) From multi-periodic cycles to chaos   Since oftentimes exact solutions are not derived, it is very difficult to know analytically what combination of nonlinear systems and parameters lead to multi-periodic solutions and chaos[171]. For example, the internal parameter (β in the previous example) that dominates the behavior of the solution varies for every interval of the independent variable. There will be a shift from dual-periodic behavior, to multi-periodic behavior.

Why do various frequencies that differ from the natural frequency of the internal system appear in the solution? Recall Figure 1.4. The linear term acts as an amplifier that increases (or decreases) the amplitude of the periodic function without affecting the frequency. The non-linear term is a frequency modulator which can generate a periodic function different from the natural frequency because of the multiplication of periodic functions.

(2) Extreme sensitivity to the initial conditions   The solution of chaos largely depends on the value of the initial condition. Slight difference in initial values results in a very large difference in behavior with time. Care should be taken, therefore, in setting up the initial value because erroneous results occur even for slight differences in the assumed initial value. It is ideal if the initial conditions (initial position and velocity) can be strictly determined, such as in the case of launching a rocket. However, in forecasting weather using ODE, the measured value of say, temperature, at a certain time is set as the initial value. Errors in the measured initial value should be considered as a factor to inaccurate forecasts.

(3) Ensemble forecast   Chaos is deterministic; such that, the solution will never change as long as the ODE representing the system and the initial condition are fixed. However, given the uncertainty of the initial value, long-term forecasts can not be trusted. In recent applications of numerics to forecasting, a method called ensemble forecasting is used. Rather than giving exact value forecasts, probability of occurrence is reported. The simulation is conducted multiple times considering multiple initial values within the range of uncertainty. After which, the deterministic values are statistically analyzed to determine the likelihood or probability of the future behavior of the dependent variable[172].

(4) Chaos emerges from the non-linear internal system   If a certain physical quantity shows an irregular behavior, it is common to think that this arises from the external irregular fluctuations induced towards the system. However, even if the external system is a simple periodic function or a constant, irregular fluctuations such as chaos arises mainly from the non-linear interaction within the internal system. In a linear system, the fluctuation component that constitutes the solution, whether a two-period solution, wave, or resonance, corresponds to the sum of the terms comprising the natural frequency of the system and the frequency of the external force. In chaotic systems, fluctuations of various frequency other than the natural frequency and the frequency of the external force appears.

(5) Strange attractor   It would be surprising even for non-engineering majors to see that the solution of a simple non-linear ODE when plotted

> **171**   Exact solutions cannot be derived for most non-linear ODEs. Numerics is mostly relied upon to solving them. It was initially suspected that chaos represents the error in numerics. It took some time before chaos was acknowledged as a physical phenomenon.

> **172**   Consider weather forecasting tomorrow's temperature using a differential equation. Assuming an observed temperature during initial time contains an error, multiple simultaneous calculations are done while slightly shifting the initial temperature values. For example, 100 calculations could be done resulting in 40 cases having 30°C or lower values. In this condition, reported weather forecast will be something similar to stating "there is a 60% that tomorrow's temperature will be over 30°C."

on a phase space results in an orbiting fractal pattern. The behavior does not move in a random manner but rather follows a rule that can be defined only by identifying the strange attractor. Interestingly, visualizing strange attractors can also be considered art (a.k.a. fractal art).

6.3 Characteristics of chaos in the non-linear mass–spring system

Let us investigate the chaotic behavior of the non-linear mass–spring system (eq. 6.3) by constructing a Python script. By now, the reader is already accustomed to the basic structure of a Python script. Refer to the previous exercises for details on its construction.

First, import the necessary modules.

```
1  import numpy as np
2  import matplotlib.pyplot as plt
```

Implementing the 4th order Runge–Kutta to minimize truncation errors, the following functions are added to numerically solve for eq. 6.3 using its modified form shown in eq. 6.4.

```
1   def runge(dt,vinit,xinit,beta,wp,Fp,w,tstart=0.,
       tend=100.):
2           t = np.arange(tstart,tend+dt,dt)
3           V = np.copy(t)
4           X = np.copy(t)
5           V[0] = vinit
6           X[0] = xinit
7           for i in range(0,t.shape[0]-1,1):
8                   h1 = dt*g(t[i]        ,V[i]       ,
                       X[i]        ,beta,wp,Fp,w)
9                   k1 = dt*h(t[i]        ,V[i]       ,
                       X[i]        ,beta,wp,Fp,w)
10                  h2 = dt*g(t[i]+0.5*dt,V[i]+0.5*h1,
                       X[i]+0.5*k1,beta,wp,Fp,w)
11                  k2 = dt*h(t[i]+0.5*dt,V[i]+0.5*h1,
                       X[i]+0.5*k1,beta,wp,Fp,w)
12                  h3 = dt*g(t[i]+0.5*dt,V[i]+0.5*h2,
                       X[i]+0.5*k2,beta,wp,Fp,w)
13                  k3 = dt*h(t[i]+0.5*dt,V[i]+0.5*h2,
                       X[i]+0.5*k2,beta,wp,Fp,w)
14                  h4 = dt*g(t[i]+dt     ,V[i]+h3    ,
                       X[i]+k3     ,beta,wp,Fp,w)
15                  k4 = dt*h(t[i]+dt     ,V[i]+h3    ,
                       X[i]+k3     ,beta,wp,Fp,w)
16                  V[i+1] = V[i]+(h1+2*h2+2*h3+h4)/6.
17                  X[i+1] = X[i]+(k1+2*k2+2*k3+k4)/6.
18          return t,X,V
19
20  def g(t,V,X,beta,wp,Fp,w):
```

```
21              return -1.*(wp*wp)*X-beta*wp*wp*X**3.+Fp*
                np.sin(w*t)
22
23  def h(t,V,X,beta,wp,Fp,w):
24          return V
```

Looking at the defined function for Runge–Kutta ("runge"), the required arguments are "dt," "vinit," "xinit," "beta," "wp," "Fp," "w," "tstart," and "tend." "dt" corresponds to the time-step. "vinit" and "xinit" are the values for V and X at the initial time 0 s, respectively. "beta" is β. "wp," "Fp," and "w" represents ω_p, F_p, and ω, respectively. "tstart" and "tend" represent the start time and end time of investigation. For convenience, they are set initially as follows. Bear in mind that the physical units are assumed to be the same as in the previous mass–spring set-ups.

```
1   dt = 0.001
2   xinit = 0.
3   vinit = 0.
4   wp = 1.*np.pi
5   Fp = 100.
6   w = 2.*np.pi
7   beta = 100.
```

After coding the module importation, the functions to solve the non-linear mass spring system, and initializing the required arguments, we can now investigate the characteristics of chaos.

6.3.1   Multi-periodic cycles

The first obvious characteristic for chaotic systems can be seen from the time-series plot of the displacement X. To obtain the time-series plot, append the following script which contains the line for calling the solution, and visualizing the time-series.

```
1   val = runge(dt,vinit,xinit,beta,wp,Fp,w,tstart=0.,
       tend=100.)
2   #note: val contains 3 values corresponding to:
3   #            t (val[0]), X(val[1]), and V(val[2])
4
5   fig = plt.figure(figsize=(18,6))
6   plt.plot(val[0],val[1])
7   plt.xlim([0,20])
8   plt.xlabel('Time␣(s)')
9   plt.ylabel('Displacement␣(m)')
10  plt.show()
```

Multiple values of periods and amplitudes can be seen of indistinguishable patterns after running the program (Figure 6.2).

6.3.2   Extreme sensitivity to the initial conditions

What will happen to the system if we modify the initial condition $X(0)$

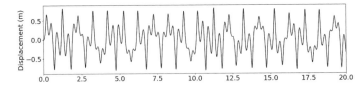

Figure 6.2 Time series of the displacement of a non-linear mass–spring system

from 0.0 m to 0.01 m? To investigate this, a time series is added to Figure 6.2. This new time series is generated when the updated initial condition $X(0) = 0.01$ is introduced to the numerical model. To achieve this, the earlier code is replaced with the following script. In the following script, carefully notice the variations in the 2nd argument used in calling the "runge" function.

```
1  val = runge(dt,vinit,xinit,beta,wp,Fp,w,tstart=0.,
     tend=100.)
2  val1 = runge(dt,vinit,0.01,beta,wp,Fp,w,tstart=0.,
     tend=100.)
3  #Notice xinit is replaced with 0.01 for
     calculating val1.
4
5  fig = plt.figure(figsize=(18,4))
6  plt.plot(val[0],val[1])
7  plt.plot(val1[0],val1[1])
8  plt.xlim([0,20])
9  plt.xlabel('Time (s)')
10 plt.ylabel('Displacement (m)')
11 plt.show()
```

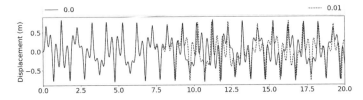

Figure 6.3 Time series of the displacement of a non-linear mass–spring system

Figure 6.3 demonstrates that a slight change in the initial condition will result in significant differences in the behavior of the system. Replacing the initial condition from 0.0 m to 0.01 m, will cause the system to behave differently in less than 2 s since initialization. Even setting the initial condition of the displacement to 0.001 m also results in significant differences. Feel free to test various initial conditions.

In real-world applications, initial conditions are measured. All measuring devices have varying levels of precision. Most of which are not precise enough to minimize the extreme sensitivity of chaotic systems. Weather forecasting is a good example for demonstrating chaos. Weather models rely on actual measurements to set the initial conditions. Not only the precision of the measurements but also the lack of measurements across the globe (resulting in initial values arising from interpolation) is the reason why weather forecasts become unreliable for longer periods (e.g. weeks) ahead. Nevertheless, this does not mean weather forecasts or forecasts of chaotic systems are useless. **Ensemble averaging** may be implemented to models of chaotic systems (considering the precision of observations) to provide probable range of forecasts. What other chaotic systems can you think of? Appreciate the fact that those systems are highly sensitive to the initial conditions.

6.3.3 Chaos comes from non-linear internal systems

Chaos arises from the non-linear term introduced in the ODE (eq. 4.10). The internal parameter responsible for chaos to emerge in the mass–spring system is β which largely influences the capacity of the spring to restore the mass point to the position of zero displacement. Let us test three values of β, namely: 0, 10, and 100. To accomplish this sensitivity test, the earlier code may be replaced with the following script. In the following script, carefully notice the variations in the 4th argument used in calling the "runge" function.

```
1   val = runge(dt,vinit,xinit,beta,wp,Fp,w,tstart=0.,
        tend=100.)
2   val1 = runge(dt,vinit,xinit,10.,wp,Fp,w,tstart=0.,
        tend=100.)
3   val2 = runge(dt,vinit,xinit,0.,wp,Fp,w,tstart=0.,
        tend=100.)
4   #Notice beta is replaced for calculating val1 and
        val2.
5
6   fig = plt.figure(figsize=(18,4))
7   plt.plot(val[0],val[1],label='0')
8   plt.plot(val1[0],val1[1],label='10')
9   plt.plot(val2[0],val2[1],label='100')
10
11  plt.xlim([0,20])
12  plt.xlabel('Time␣(s)')
13  plt.ylabel('Displacement␣(m)')
14  plt.legend()
15  plt.show()
```

Figure 6.4 shows the influence of the internal parameter β to the mass spring system. When β is set to 0.0, harmonic oscillation occurs since the non-linear ODE in eq. 6.2 becomes a 2nd order linear non-homogeneous ODE (recall previous chapter). However, changing the value of β drastically increases the spring coefficient at various displacement values. At $\beta = 10$, the amplitude of the displacement is drastically reduced. In this condition (check by showing only the result of $\beta = 10$), however, a fundamental period can still be inferred from the resulting time series unlike the condition when $\beta = 100$ (Figure 6.2).

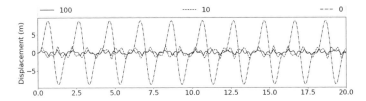

Figure 6.4 Time series of the displacement of a non-linear mass–spring
system. Note: $X(0) = 0.0$

6.3.4 Strange attractor in the non-linear mass–spring system

As earlier mentioned, a fascinating characteristic of chaos is called
"strange attractors" which is also hidden in the non-linear mass–spring system. To visualize strange attractors, we will generate sequences of Poincaré
sections by iteration. A Poincaré section is a phase diagram (X vs. V) of
displacement samples taken at $\omega/2\pi$ s (equivalent to 1.0 s from the set-up
above) equal time intervals. For example, X and V at 0.0 s., 1.0 s, 2.0 s,...
may be plotted as one Poincaré section. Another section may be plotted
from samples taken at 0.0 s + δt, 1.0 s + δt, 2.0 s + δt, and so on; or at
0.0 s + $2.0\delta t$, 1.0 s + $2.0\delta t$, 2.0 s + $2.0\delta t$, and so on. Below is a script to
generate sequences of Poincaré sections and storing them as separate im-
ages. A time-lapse software may be used to create an animation of Poincaré
sections to reveal the strange attractors.

```
1   val = runge(dt,vinit,xinit,beta,wp,Fp,w,tstart=0.,
       tend=10000.)
2   #Notice we extend the tend to 10000. s to acquire
       more samples.
3
4   for f in range(0,1000):
5           fig = plt.figure(f)
6           plt.scatter(val[1][range(f,val[1].shape
               [0],1000)],\
7                           val[2][range(f,val[1].
                               shape[0],1000)],\
8                           c=val[0][range(f,val[1].
                               shape[0],1000)],\
9                           vmin=0.,vmax=10000.,s
                               =2.)
10          plt.xlim([-1.,1.])
11          plt.ylim([-20.,20.])
12          plt.xlabel('Displacement␣m')
13          plt.ylabel('Velocity␣m/s')
14          plt.colorbar()
15          plt.savefig('./result_%05d.png'%(f),
               bbox_inches='tight')
16          #The output will be multple 'png' files
               with filenames
17          #starting in 'result..'
18          plt.close(fig)
```

Figure 6.5 shows a few selected Poincaré sections (from the program
they are result_00000.png, result_00250.png, result_00500.png, and re-

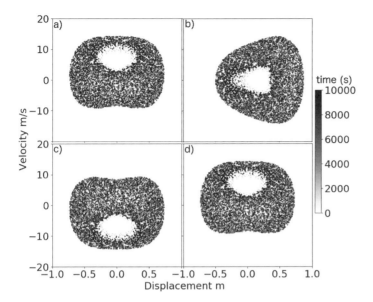

Figure 6.5 Poincaré sections taken at various starting positions. a, b, c, and d were taken from 0.0 s (i.e. samples include X and V values at 0.0, 1.0, 2.0,...,10000 seconds), 0.25 s (i.e. samples include X and V values at 0.25, 1.25, 2.25,..., 999.25 seconds), 0.5 s (i.e. samples include X and V values at 0.5, 1.5, 2.5,..., 999.5 seconds), and 0.999 s (i.e. samples include X and V values at 0.999, 1.999, 2.999,..., 999.999 seconds), respectively.

sult_00999.png). It can be seen that the samples representing X and V at various times equally spaced at $\omega/2\pi$ s are drawn (or attracted) towards a specific region in the phase diagram. This is one evidence that chaos is not a random phenomenon. In other words, chaotic systems remain subjected to laws hidden within the system defined by ODEs (or PDEs if independent variables are more than 1). It is also interesting that the Poincaré section taken from 0.999 s is almost similar to that taken from 0.0 s. From Figure 6.5 or by observing all generated images in sequence, it can be seen that all Poincaré sections are interrelated, and loop as it approaches a full period determined by the external angular frequency; thereby, not departing from a region of attraction. Our world is filled with chaotic systems such as the non-linear mass–spring systems. Isn't it interesting to try to imagine the strange attractors within each of them?

6.4 Food chain and chaos

6.4.1 Population of predator and pray: Lotka–Volterra equation

As an additional example of chaos arising from non-linear phenomena, biological growth already discussed in ch. 2 will be reviewed.

Exponential growth/decay: Recall eq. 2.3 in ch. 2 while assuming an absent external system.

$$\frac{dN}{dt} = aN$$

Saturated growth...Logistic curve: Recall eq. 71 in ch. 2.

$$\frac{dN}{dt} = (a - bN)N$$

Predator/prey growth...Lotka–Volterra equation: Current chapter

$$\frac{dN_1}{dt} = (a - bN_2)N_1,$$

$$\frac{dN_2}{dt} = (-c + dN_1)N_2. \tag{6.5}$$

In the case of simple biological growth, such as cell division which does not experience environmental suppression, the growth rate of organisms will be proportional to their population; so, it is often acceptable to set a constant self-propagation rate a (e.g. attenuation condition in eq. 2.11). On the other hand, competition for food and deterioration of the environment may occur due to an increase in the number of individuals, thus suppressing the growth rate. In this case, the self-propagation rate is assumed to be a linearly decreasing function $(a - bN)$ of the number of individuals. The solution is a logistic curve. However, let us assume two species existing in the same environment where one specie preys on another. N_1 is an organism (prey) that can be eaten by another organism N_2 (predator). In other words, N_2 feeds on N_1. Consider the model in eq. 6.5. The predator N_1 has a constant self-propagating function a if there is no predator N_2. However, N_2 exists and is also feeding on the prey; thus, N_1's survival is proportional to the population of N_1 and N_2. On the other hand, the self-propagation rate of the predator N_2, which appears at the top of the food chain, is negative $(-c)$ as its population decreases due to starvation and death. As a lone predator, N_2 will increase proportional to the number of individuals N_1 by $(-c + dN_1)N_2$. This simple non-linear predator-prey model is known as Lotka–Volterra and is written in eq. 6.5. Since the equation is a 1st order non-linear ODE, exact solutions are not easily derived; thus, the behavior of the solutions are analyzed using the numerical solution, Runge–Kutta method.

6.4.2 Subordinates (prey), mid-rank organisms (predator-prey), top organism (predator): Chaos phenomenon from co-existence of 3 organisms types

Let us extend the idea of the Lotka–Volterra equation of only one predator to a biological pyramid consisting of three levels. Subordinate organisms N_1 (prey), mid-rank organisms N_2 (both predator and prey), and superior organisms N_3 (predator) are set. Consider a simple food chain model described by the simultaneous ODEs below (eq. 6.6). The conceptual diagram is shown in Figure 6.6. The simultaneous ODEs describe the time rate of change of the population density N_1, N_2, N_3 (number per km^2) of the the individual species.

$$\frac{dN_1}{dt} = (b_1 - a_{11}N_1 - a_{12}N_2 - a_{13}N_3)N_1,$$

$$\frac{dN_2}{dt} = (-b_2 + a_{21}N_1 - a_{23}N_3)N_2,$$

$$\frac{dN_3}{dt} = (-b_3 + a_{31}N_1 + a_{32}N_2)N_3. \tag{6.6}$$

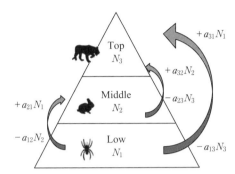

Figure 6.6　Conceptual diagram of a food chain model composed of 3 kinds of organisms.

First, consider the self-growth rate of subordinate organisms. The subordinate organisms N_1, which is the prey or the basal resource, is consumed by both mid-rank organisms N_2 and the superior organisms N_3. If predators N_2 and N_3 are absent, a logistic type of self-propagation rate $(b_1 - a_{11}N_1)$ is assumed which means growth is suppressed according to the population density of N_1. N_1 is a prey for both N_2 and N_3, thus additional terms that suppress the propagation of N_1 is introduced. This is represented by a proportion of the population density of the predators $(b_1 - a_{11}N_1 - a_{12}N_2 - a_{13}N_3)$. On the other hand, the self-propagation rate of the superior (top) organisms N_3 at the top of the food chain is a negative value $(-b_3)$ since the population density decreases as a result of starvation and death in the absence of low ranking creatures N_1 and N_2. Self-propagation is achieved by hunting and consuming preys. Its self-propagation rate is proportional to the number of N_1 and N_2 $(-b_3 + a_{31}N_1 + a_{32}N_2)$. The mid-rank organisms N_2 is a predator of the subordinate organism N_1 and also a prey of the superior organism N_3. If N_1 and N_3 disappears, the population decreases as a result of starvation, thus the self-propagation rate is $(-b_2)$. If both N_1 and N_3 are present, its self-propagation is proportional to the population density N_1 and N_3 $(-b_2 + a_{21}N_1 - a_{23}N_3)$.

The food chain model composed of three simplified representation of organisms is comprised of 1st-order non-linear ODEs. Since exact solutions are difficult to derive, the behavior can be represented numerically using the Runge–Kutta method. Chaos does not necessarily occur from Lotka–Volterra equation comprising two interacting organisms, but instead occurs in the three-level Lotka–Volterra equation depending on the value of the internal parameters.

6.4.3 Chaos in food web models

Lotka–Volterra equation when revised to a three-organism interaction (as represented by simultaneous ODEs in eq. 6.6) are generally called food web models. Depending on the internal parameters of the system, chaos may occur. Let us investigate the condition of chaos using the 4th order Runge–Kutta model using the common setup below. As with the non-linear mass–spring investigation, the necessary modules are called and a 4th order Runge–Kutta model for eq. 6.6 is constructed. For simplicity, the internal parameters and initial conditions are predetermined (which in practice are estimated from observations) and are coded prior to the definition of the "rk" function which contains the numerical solution. This eliminates the need of inputting the predefined variables as arguments to the "rk" function. The analyses period is set to 0 to 1000 s. with a calculation time step of 0.01 s. The variables are self-explanatory.

```
1   import matplotlib.pyplot as plt
2   import numpy as np
3   from matplotlib import cm # for colormap
4   from mpl_toolkits.mplot3d import Axes3D # for
      plotting 3D
5   import peakutils # for estimating peak values
6
7   a11 = 0.4
8   a12 = 1.0
9   a13 = 7.
10  a21 = 1.0
11  a23 = 1.0
12  a32 = 1.0
13  a31 = 0.1
14  b1 = 5.0
15  b2 = 1.0
16  b3 = 1.2
17
18  N1init = 1.
19  N2init = 2.
20  N3init = 3.
21
22  delta = 0.01
23  t = np.arange(0.,1000.+delta,delta)
24
25  def rk(t,delta,N1init,N2init,N3init):
26      N1 = np.copy(t)
27      N2 = np.copy(t)
28      N3 = np.copy(t)
29      N1[0] = N1init
30      N2[0] = N2init
31      N3[0] = N3init
32      for i in range(0,t.shape[0]-1):
33          h1 = delta*dN1dt(N1[i],N2[i],N3[i])
34          k1 = delta*dN2dt(N1[i],N2[i],N3[i])
35          l1 = delta*dN3dt(N1[i],N2[i],N3[i])
36          h2 = delta*dN1dt(N1[i]+h1/2.,N2[i]+k1/2.,
              N3[i]+l1/2.)
37          k2 = delta*dN2dt(N1[i]+h1/2.,N2[i]+k1/2.,
              N3[i]+l1/2.)
38          l2 = delta*dN3dt(N1[i]+h1/2.,N2[i]+k1/2.,
              N3[i]+l1/2.)
```

```
39          h3 = delta*dN1dt(N1[i]+h2/2.,N2[i]+k2/2.,
               N3[i]+12/2.)
40          k3 = delta*dN2dt(N1[i]+h2/2.,N2[i]+k2/2.,
               N3[i]+12/2.)
41          13 = delta*dN3dt(N1[i]+h2/2.,N2[i]+k2/2.,
               N3[i]+12/2.)
42          h4 = delta*dN1dt(N1[i]+h3,N2[i]+k3,N3[i]+
               13)
43          k4 = delta*dN2dt(N1[i]+h3,N2[i]+k3,N3[i]+
               13)
44          14 = delta*dN3dt(N1[i]+h3,N2[i]+k3,N3[i]+
               13)
45          N1[i+1] = N1[i]+(h1+2.*h2+2.*h3+h4)/6.
46          N2[i+1] = N2[i]+(k1+2.*k2+2.*k3+k4)/6.
47          N3[i+1] = N3[i]+(11+2.*12+2.*13+14)/6.
48      return N1,N2,N3
49
50  def dN1dt(N1,N2,N3):
51      return (b1-a11*N1-a12*N2-a13*N3)*N1
52
53  def dN2dt(N1,N2,N3):
54      return (-b2+a21*N1-a23*N3)*N2
55
56  def dN3dt(N1,N2,N3):
57          return (-b3+a31*N1+a32*N2)*N3
```

As a first analysis, a time-series is plotted to see the changes of N_1, N_2, and N_3 with time. The following function can be appended to the above script.

```
1   def timeseries():
2           plt.figure(1,figsize=(12,6))
3           val = rk(t,delta,N1init,N2init,N3init)
4           for i in range(0,3):
5               plt.plot(t,val[i],label='N%1i'%(i+1))
6           plt.xlim([0.,100.])
7           plt.xlabel('Time␣(s)')
8           plt.ylabel('Population␣Density')
9           plt.legend(loc=0)
10          plt.show()
```

From the script above, plotting a time series is written inside the function, "timeseries()." Adding the line "timeseries()" at the bottom of the script and running the program will display Figure 6.7. Under the given internal parameters, a clear pattern could be seen in the increase and decrease of population. A fix periodic behavior is seen from all species. N_1 which is capable of exponential growth increases in population only up to a certain time (12.5 organisms per area) because it is continually being fed upon by both N_2 and N_3. The population growth of N_2 is also being affected by N_3. N_3 which feeds on N_1 and N_2 is also limited by its consumption rate and quantity of its preys. Under the stated growth and interaction rates, the population peaks and its timing (period) is constant, nonetheless.

The situation changes when one or a few internal parameters which determine the strength of non-linearity is increased. To investigate this, let us

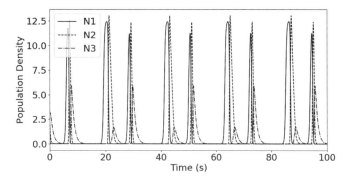

Figure 6.7   Time-series plot of N_1, N_2, and N_3

replace the value of a_{13} (a13 = 7.) with 15 (or 13 = 15.). The behavior of the system is now chaotic as seen from Figure 6.8. The species behave in a seemingly unexplainable pattern. For example, two peaks could be seen in N_2 with only one peak for N_3 in between 40 to 60 s since initialization. Furthermore, a clear periodic behavior which can be seen in Figure 6.7 can no longer be seen in this case. This further demonstrates the characteristic of chaos that it is caused by the condition of the internal parameters.

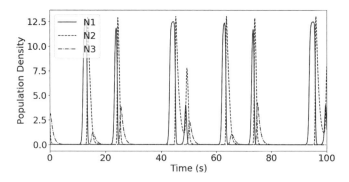

Figure 6.8   Time-series plot of N_1, N_2, and N_3 when a_{13} is set to 15

In the same chaotic set-up, the sensitivity to the initial conditions may also be analyzed using the following code. The time series for N_1, N_2, and N_3 are plotted from 900 to 1,000 s for two cases; one where the initial values are unchanged and for another where the initial value of N_1 is added with 0.001 organisms per area (may be coded Ninit = Ninit + 0.001 prior to calling the "rk" function). From Figure 6.9, it can be seen that the propagation and elimination of the species have drastically changed despite slight changes in the initial population density.

```
1   def timeseries_sensitive_to_initial():
2       plt.figure(1,figsize=(12,6))
3       global N1init
4       val = rk(t,delta,N1init,N2init,N3init)
5       for i in range(0,3):
6           plt.plot(t,val[i],label='N%1i'%(i+1))
```

```
 7
 8                #The next code are additional cases where
                      N1init is added with 0.001
 9                del val
10                val = rk(t,delta,N1init+0.001,N2init,
                      N3init)
11                for i in range(0,3):
12                    plt.plot(t,val[i],label='N%1i_b'%(i+1)
                          ,linewidth=0.5)
13
14                plt.xlim([900.,1000.])
15                plt.legend(loc=1)
16                plt.show()
```

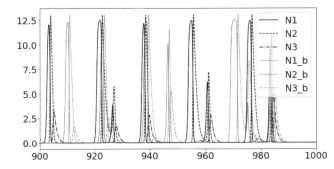

Figure 6.9   Time-series plot of N_1, N_2, and N_3 under a case when initial conditions are unchanged and a case when the initial condition for N_1 is added by 0.001. The results of the latter case are labeled "N1_b," "N2_b," and "N3_b." a_{13} was set to 15 for both cases

Strange attractors can also be visualized by a 3D scatter plot of N_1, N_2, and N_3 using the following code. Figure 6.10 shows a strange attractor for the given system. Irrespective of the time, the combinations of N_1, N_2, and N_3 are drawn towards a region in the 3D plot. There is no clear pattern when the samples are situated close or far from the region it is orbiting.

```
 1  def strangeattractor():
 2          val = rk(t,delta,N1init,N2init,N3init)
 3
 4          fig = plt.figure(figsize=(8,8))
 5          ax = fig.gca(projection='3d')
 6          surf = ax.scatter(val[0],val[1],val[2],c=t
                  ,s=1.,cmap=cm.gist_rainbow)
 7          fig.colorbar(surf, shrink=0.5, aspect=5)
 8          ax.set_xlabel('N1')
 9          ax.set_ylabel('N2')
10          ax.set_zlabel('N3')
11          ax.set_xlim([0.,12.])
12          ax.set_ylim([0.,12.])
13          ax.set_zlim([0.,6.])
14          plt.show()
```

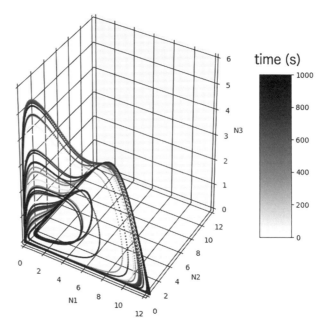

Figure 6.10 3D scatter plot of N_1, N_2, and N_3. The color bar denotes the timing of the samples.

The analyses conducted earlier follow the similar approach done for the non-linear mass–spring system. Here, an interesting chart which demonstrates the multi-periodic behavior leading to chaos is introduced. This is achieved by analyzing the sensitivity of a dependent variable's peak value to the internal parameter which triggers chaotic behavior. In the following code (be forewarned that the simulation time may be longer than usual), the value of a_{13} is iterated from 5. to 16. at 0.01 intervals, and the corresponding peak values of N_1 (corresponding to val[2] on the code below) are collected from a certain time until 5000 s and plotted. The multiple peak values throughout the time series are extracted using the Python library, "peakutils."

The phenomenon which can be seen from the resulting plot (Figure 6.11) is "bifurcation." As the value of a_{13} increases, a fork in the maximas appear. With further increases in a_{13}, a similar fork is reproduced with a scale finer than the previous one in both top and lower branches of the earlier fork. This continues infinitely with increasing a_{13}, which resembles fractals. The forks also represents period-doubling. At more chaotic conditions (large a_{13} values), multiple peaks or periods are revealed because of the occurrence of bifurcations.

```
1   def bifurcation():
2       plt.clf()
3       fig = plt.figure(1,figsize=(8,8))
4       delta = 0.05
5       t = np.arange(0.,5000.+delta,delta)
6       global a13
7       for a13 in np.arange(5.,16.,0.005):
8           val = rk(t,delta,N1init,N2init,
```

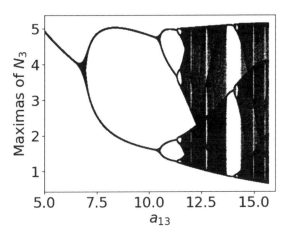

Figure 6.11 Bifurcation diagram of peaks of N_3 under varying a_{13} values.

```
                    N3init)[:]
9               ind = peakutils.indexes(val
                [2][10000:100000],\
10                    thres = 0.,min_dist=1)
11              x = np.empty(ind.shape[0],dtype=np
                .float32)
12              x[:] = a13
13              plt.plot(x,val[2][10000:100000][
                ind],'k.',markersize=1.)
14          plt.xlim('a13')
15          plt.ylim('Maximas␣of␣N3')
16          plt.show()
```

References used in this section:

• Narendra S. Goel, Samaresh C. Maitra, and Elliott W. Montroll. On the Volterra and other nonlinear models of interacting populations. *Reviews of Modern Physics*, 43(2):231–276, 1971.

• Kumi Tanabe, and Toshiyuki Namba. Omnivory creates chaos in simple food web models. *Ecology*, 86(12):3411–3414, 2005.

6.5   Exercise 5

6.5.1   Lorenz model

As an exercise for chaos in ODE, it is appropriate to investigate the Lorenz model which can be summarized by the following simultaneous ODEs:

$$\frac{dX}{dt} = -\sigma X + \sigma Y$$

$$\frac{dY}{dt} = -XZ + rX - Y$$

$$\frac{dZ}{dt} = XY - bZ$$

X, Y, and Z are non-dimensional dependent variables which represent convective intensity, temperature difference between descending and ascending currents, and the difference in the vertical temperature profile, respectively. The internal parameters σ, r, and b, represents the Prandtl number, Rayleigh number, and geometric factor, respectively. It is beyond the scope of this textbook to discuss these parameters in detail. However, by analyzing the equations above mathematically through numerics (4th order Runge–Kutta method), let us investigate chaos phenomena by conducting the following analyses. Assume $\sigma = 10.$, $r = 28.$, and $b = 8./3.$, and initial condition of $X(t=0) = X_0 = 1.$, $Y(t=0) = Y_0 = 1.$, $Z(t=0) = Z_0 = 1.$ (Dirichlet boundary condition).

1. Plot the time-series of X, Y, and Z from time $t = 0$ to $100s$ with a step size of $0.01s$.

2. Superimpose a time-series of Z to the plot in (1) using a different intial condition for $X_0 = 1.5$.

3. Construct a 3D plot X,Y, and Z using different initial conditions of X_0 (1.,1.5,10.).

Reference to the Lorenz model:

• Edward N. Lorenz. Deterministic nonperiodic flow. *Journal of the Atmospheric Sciences*, 20(2):130–141, 1963.

6.5.2 Procedures for solving Exercise 5

The importing of necessary modules, the assignment of constant parameters, and the definition of the function and gradients are written below.

```
1   #Importing modules
2   import numpy as np
3   import matplotlib.pyplot as plt
4   import matplotlib as mpl
5   from mpl_toolkits.mplot3d import Axes3D
6
7   #Definition of constant variables
8   sigma = 10.
9   r = 28.
10  b = 8./3.
11
12  #Runge-Kutta function definition.
13  def rk(delta,t,xinit,yinit,zinit):
14          x = np.copy(t)
15          y = np.copy(t)
16          z = np.copy(t)
17          x[0] = xinit
18          y[0] = yinit
19          z[0] = zinit
```

```
20              for i in range(0,t.shape[0]-1):
21                      h1 = delta*dxdt(x[i],y[i])
22                      k1 = delta*dydt(x[i],y[i],z[i])
23                      j1 = delta*dzdt(x[i],y[i],z[i])
24                      h2 = delta*dxdt(x[i]+h1/2.,y[i]+k1
                            /2.)
25                      k2 = delta*dydt(x[i]+h1/2.,y[i]+k1
                            /2.,z[i]+j1/2.)
26                      j2 = delta*dzdt(x[i]+h1/2.,y[i]+k1
                            /2.,z[i]+j1/2.)
27                      h3 = delta*dxdt(x[i]+h2/2.,y[i]+k2
                            /2.)
28                      k3 = delta*dydt(x[i]+h2/2.,y[i]+k2
                            /2.,z[i]+j2/2.)
29                      j3 = delta*dzdt(x[i]+h2/2.,y[i]+k2
                            /2.,z[i]+j2/2.)
30                      h4 = delta*dxdt(x[i]+h3,y[i]+k3)
31                      k4 = delta*dydt(x[i]+h3,y[i]+k3,z[
                            i]+j3)
32                      j4 = delta*dzdt(x[i]+h3,y[i]+k3,z[
                            i]+j3)
33                      x[i+1] = x[i] + (h1+2.*h2+2.*h3+h4
                            )/6.
34                      y[i+1] = y[i] + (k1+2.*k2+2.*k3+k4
                            )/6.
35                      z[i+1] = z[i] + (j1+2.*j2+2.*j3+j4
                            )/6.
36              return x,y,z
37
38      #Defining the function for the gradient.
39      def dxdt(x,y):
40              return sigma*(y-x)
41      def dydt(x,y,z):
42              return x*(r-z)-y
43      def dzdt(x,y,z):
44              return x*y-b*z
```

To obtain the required time-series plots, the program below may be implemented.

```
1   delta = 0.01
2   xinit = 1.
3   yinit = 1.
4   zinit = 1.
5   t = np.arange(0,100.+delta,delta)
6   x,y,z = rk(delta,t,xinit,yinit,zinit)
7   fig = plt.figure(1)
8   plt.plot(t,x,label='x')
9   plt.plot(t,y,label='y')
10  plt.plot(t,z,label='z')
11  plt.legend(loc=0)
12  plt.show()
```

Running the program, will result in the following figure.

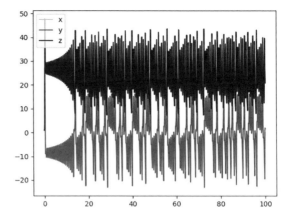

The program above may be improved to see the effects of modifications in the initial condition (e.g. $X_0 = 1.5$) by recalculating the time-series of x, y, and z under the updated initial condition. As already discussed, the system is chaotic in the sense that it is highly sensitive to the initial condition. Feel free to modify the initial conditions and see how sensitive it is.

To construct a 3D plot for X, Y, and Z (requirement no. 3), the following code may be implemented.

```
1  fig = plt.figure(2)
2  ax = fig.gca(projection='3d')
3  ax.plot(x1, y, z, label='Lorenz␣Attractor␣1',c='
     black')
4  x2,y,z = rk(delta,t,xinit+0.5,yinit+0.5,zinit+0.5)
5  ax.plot(x2, y, z, label='Lorenz␣Attractor␣2',c='
     red')
6  x3,y,z = rk(delta,t,xinit+10.,yinit+100.,zinit
     +100.)
7  ax.plot(x3, y, z, label='Lorenz␣Attractor␣3',c='
     green')
8  ax.legend()
9  plt.show()
```

Running the program above, will result in the following figure.

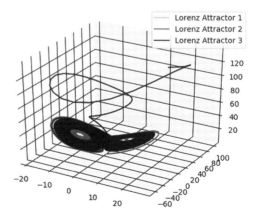

Regardless of the initial condition, the result shows a strange attractor (Lorenz attractor) that resembles the shape of a butterfly (does "butterfly effect" ring a bell?).

◆◆◆◆◆◆ Problems for Exercise ◆◆◆◆◆◆

6.1   Lorenz attractor

A phase diagram of the following non-linear ODE's solution may display a typical example of a strange attractor called the Lorenz attractor. Reproduce it using the 4th order Runge–Kutta method with a time-step of 0.01 (any time unit) for a maximum time duration of 30. The internal parameters are σ, γ, and b with values of 10, 28, and 8/3, respectively. Consider two sets of initial values for X_0, Y_0, and Z_0 to be (1.0, 2.0, 10.0) and (1.1, 2.0, 10.0). Construct (1) time-series plots for X and Z, (2) X-Y phase diagrams, and (3) Y-Z phase diagrams for each sets of initial conditions.

$$\frac{dX}{dt} = -\sigma(X - Y)$$

$$\frac{dY}{dt} = -XZ + \gamma X - Y$$

$$\frac{dZ}{dt} = XY - bZ$$

6.2   Rössler attractor

A phase diagram of the following non-linear ODE's solution may produce a typical example of a strange attractor called the Rössler attractor. Reproduce it using the 4th order Runge–Kutta method with a time-step of 0.01 (any time unit) for a maximum time duration of 200. Consider initial values for X_0, Y_0, and Z_0 to be 1.0, 1.0, 1.0, respectively. The internal parameters a and b are both set to 0.1. Test 3 possible values for c (4.0, 12.0, 18.0). For each value of c, construct (1) time-series plots for X and Z, (2) X-Y phase diagrams, and (3) Y-Z phase diagrams.

$$\frac{dX}{dt} = -Y - Z$$

$$\frac{dY}{dt} = X + aY$$

$$\frac{dZ}{dt} = b + Z(X - c)$$

Answers to Problems

1

1.1 Ordinary differential equations: $1, 3, 4, 6$
Partial differential equations: $2, 5, 7$

1.2 Answer the following questions regarding the ODEs determined from the previous problem.

1 1:

    (a) Two initial conditions (since the ODE is of 2nd order from T_{tt})

    (b) Linear

    (c) Non-homogeneous ($\sin t$ is of zero order)

2 3:

    (a) Two boundary conditions (since the ODE is of 2nd order from T_{xx})

    (b) Non-linear ($T T_{xx}$)

    (c) Non-homogeneous (since the orders are different for each term)

3 4:

    (a) Two boundary conditions (since the ODE is of 2nd order from T_{xx})

    (b) Linear

    (c) Homogeneous (all terms are of 1st order)

4 6:

    (a) Two boundary conditions each for T and C.

    (b) Non-linear (CT)

    (c) Non-homogeneous

2

2.1

1 linear: $T = t \ln |t| - t + C$

$$\int dT = \int \ln t \, dt, \qquad T = \int (t)' \ln t \, dt = t \ln |t| - t + C.$$

2 linear: $T = \dfrac{1}{6} \ln \left| \dfrac{t-3}{t+3} \right| + C$

$$\int dT = \int \frac{1}{t^2 - 9} dt,$$

$$T = \int \frac{1}{6} \left(\frac{1}{t-3} - \frac{1}{t+3} \right) dt = \frac{1}{6} \left(\ln |t-3| - \ln |t+3| \right) + C.$$

3 linear: $T = C \exp \left(\dfrac{1}{3} t^3 \right)$

$$\int \frac{1}{T} dT = \int t^2 dt, \quad \ln|T| + C_1 = \frac{1}{3}t^3.$$

4 non-linear but solvable with variable separation method:

$$T = 2\frac{(\exp(-4/t) - C)}{(\exp(-4/t) + C)}$$

$$\int \frac{1}{4 - T^2} dT = \int \frac{1}{t^2} dt, \quad \int \frac{1}{4}\left(\frac{1}{2-T} + \frac{1}{2+T}\right) dT = \frac{-1}{t}.$$

$$\ln C_1 \left|\frac{2+T}{2-T}\right| = \frac{-4}{t}, \quad \left(\frac{2+T}{2-T}\right) = \frac{1}{C}\exp\left(\frac{-4}{t}\right).$$

5 linear: $T = t\sin t + \cos t + C$

$$\int dT = \int t(\sin t)' dt = t\sin t - \int \sin t = t\sin t + \cos t + C.$$

<hr>

2.2

1 Answer: $T = t(3\ln|t| + C)$

Substituting $u = T/t$ and $dT = tdu + udt$ to the equation,

$t\,du/dt + u = u + 3$ can be solved using variable separation method which results in,

$$u = 3\ln|t| + C$$

2 Answer: $T = t\sqrt{4\ln|t| + C}$

Substituting $u = T/t$ and $dT = tdu + udt$ to the equation,

$t\,du/dt = 2/u$ can be solved using variable separation method which results in,

$$u^2 = 4\ln|t| + C$$

<hr>

2.3   Answer: $v = \dfrac{\exp(19.6t) - 1}{\exp(19.6t) + 1}$

$$\frac{dv}{dt} = 9.8(1 - v^2)^{\circ\mathbf{173}}, \quad \ln C_1 \left|\frac{1+v}{1-v}\right| = 2 \times 9.8^{\circ\mathbf{174}}, \quad \frac{1+v}{1-v} = C\exp(19.6t).$$

The terminal velocity is 1 m/s.

⊅**173** $mdv/dt = -kv^2 + mg$, $dv/dt = -(k/m)v^2 + g$.

⊅**174** $\int 1/(1 - v^2)dv = \int 9.8dt$ $1/(1 - v^2) = (1/2)(1/(1 + v) + 1/(1 - v))$

2.4   Answer: $v = -\exp(-9.8t) + 1$

The terminal velocity is the same as the previous problem's.

2.5   Let us apply eq. 2.23 and eq. 2.24 of this chapter. For this problem, the control volume in the box model is represented by the entire body of each species, where its body temperature is the dependent variable C. It is assumed that there is no difference between dinosaurs and small animals in terms of heat exchange coefficient K and outside temperature C_{out} between the animal and the outside air[175]. Therefore, only the surface area-to-volume ratio influences the difference in time-rate of change of body temperature for both species. Since volume is proportional to the cube of a body length L and the surface area is proportional to its square, the surface area-to-volume ratio will be inversely proportional to L. This suggests that

⊅**175**   Some animals can actually adapt by changing the thickness or color of their fur.

the larger animal takes longer to emit heat from its body than smaller animals. Thus, body size is crucial for heat storage, especially during cold seasons.

2.6 Around 23 m as suggested by the following equation derived from eq. 2.7,

$$I/I_0 = 10(\%)/100 = \exp(-0.1z)$$

2.7 Answer: $\lambda = \dfrac{\ln N_2/N_1}{t_2 - t_1}$

From eq. 2.17, $N_1 = N_0 \exp(-\lambda t_1)$ $N_2 = N_0 \exp(-\lambda t_2)$

$$N_1/N_2 = \exp[-\lambda(t_1 - t_2)]$$

3

3.1

1 Answer: $T = 3e^{2t} - 2te^{2t}$

$$\lambda^2 - 4\lambda + 4 = 0, \quad \lambda = 2, \quad T = C_1 e^{2t} + C_2 t e^{2t} \circ\mathbf{176}.$$

⟳**176** $T' = (2C_1 + C_2)e^{2t} + 2C_2 t e^{2t}$
$T(0) = C_1 = 3$
$T'(0) = 2C_1 + C_2 = 4$

2 Answer: $T = -2e^{2t} + 4e^t$

$$\lambda^2 - 3\lambda + 2 = 0, \quad \lambda = 2, 1, \quad T = C_1 e^{2t} + C_2 e^t \circ\mathbf{177}.$$

⟳**177** $T' = 2C_1 e^{2t} + C_2 e^t$
$T(0) = C_1 + C_2 = 2$
$T'(0) = 2C_1 + C_2 = 0$

3 Answer: $T = e^{-2t}(2\cos t + 4\sin t)$

$$\lambda^2 + 4\lambda + 5 = 0, \quad \lambda = -2 \pm i, \quad p = -2, q = 1,$$
$$T = e^{-2t}(C_1 \cos t + C_2 \sin t) \circ\mathbf{178}.$$

⟳**178**
$T' = -2e^{-2t}(C_1 \cos t + C_2 \sin t)$
$+ e^{-2t}(-C_1 \sin qt + C_2 \cos t)$
$T(0) = C_1 = 2$
$T'(0) = -2C_1 + C_2 = 0$

4 Answer: $T = -e^{-3t}$

$$\lambda^2 + 6\lambda + 9 = 0, \quad \lambda = -3, \quad T = C_1 e^{-3t} + C_2 t e^{-3t} \circ\mathbf{179}.$$

⟳**179**
$T' = (-3C_1 + C_2)e^{-3t} - 3C_2 t e^{-3t}$
$T(0) = C_1 = -1$
$T'(0) = -3C_1 + C_2 = 3$

5 Answer: $T = \dfrac{4}{5}e^{3t} + \dfrac{1}{5}e^{-2t}$

$$\lambda^2 - \lambda - 6 = 0, \quad \lambda = 3, -2, \quad T = C_1 e^{3t} + C_2 e^{-2t} \circ\mathbf{180}.$$

⟳**180** $T' = 3C_1 e^{3t} - 2C_2 e^{-2t}$
$T(0) = C_1 + C_2 = 1$
$T'(0) = 3C_1 - 2C_2 = 2$

6 Answer: $T = e^{-t}\left(\cos 2t - \dfrac{3}{2}\sin 2t\right)$

$$\lambda^2 + 2\lambda + 5 = 0, \quad \lambda = -1 \pm 2i, \quad T = e^{-t}(C_1 \cos 2t + C_2 \sin 2t) \circ\mathbf{181}.$$

⟳**181**
$T' = -e^{-t}(C_1 \cos 2t + C_2 \sin 2t)$
$+ e^{-t}(-2C_1 \sin 2t + 2C_2 \cos 2t)$
$T(0) = C_1 = 1$
$T'(0) = -C_1 + 2C_2 = -4$

3.2

1 Answer: $T = 2t \ln|t| + C_1 t + C_2 \circ\mathbf{182}$ Linear Non-homogeneous

2 Answer: $T = 3t + C_2 - \ln|1 - C_1 e^{6t}|$ Non-linear Non-homogeneous

⟳**182** $T' = 2\ln|t| + C'$
$T = 2\int (t)' \ln t \, dt + C''t + C_2$
$= 2t \ln|t| - 2t + C''t + C_2$
$= 2t \ln|t| + C_1 t + C_2$

$$\frac{dX}{dt} = X^2 - 9, \quad \int \frac{1}{6}\left(\frac{-1}{X+3} + \frac{1}{X-3}\right)dX = \int dt, \quad \frac{X-3}{X+3} = C_1 e^{6t}$$

$$X = \frac{dT}{dt} = 3\frac{1 + C_1 e^{6t}}{1 - C_1 e^{6t}} \quad x = e^{6t} \text{ can be integrated}^{\circlearrowright\textbf{183}}.$$

$$T = \int \frac{1}{2}\frac{1 + C_1 x}{(1 - C_1 x)x}dx = \int \left(\frac{1}{2x} + \frac{C_1}{1 - C_1 x}\right)dx = \frac{1}{2}\ln x + C_2 - \ln|1 - C_1 x|$$

$\circlearrowright\textbf{183}$ $dx = 6e^{6t}dt = 6xdt$
$\int dT = \int (3(1 + C_1 x)/(1 - C_1 x))$
$(1/6x)\,dx$

3 Answer: $T = -\dfrac{1}{C}\tan^{-1}\dfrac{t}{C}$ or $T = -\dfrac{1}{2C}\ln\left|\dfrac{t-C}{t+C}\right|$

$$\frac{1}{2t}\frac{dX}{dt} - X^2 = 0, \quad \int \frac{1}{X^2}dX = \int 2t\,dt, \quad X = \frac{dT}{dt} = \frac{-1}{t^2 + C'}{}^{\circlearrowright\textbf{184}}$$

$$\int dT = \int \frac{-1}{t^2 + C^2}dt \text{ or } \int dT = \int \frac{-1}{t^2 - C^2}dt$$

$\circlearrowright\textbf{184}$ If $C' > 0$, set $C^2 = C'$
If $C' < 0$, set $C^2 = -C'$

3.3

1 Here, we utilize Newton's law. Specifically, the conservation of angular momentum states that the sum of the external torque is equal to the moment of inertia multiplied by its angular acceleration. Mathematically, it can be represented by the following,

$$M\frac{d^2\theta}{dt^2} = N = r \times F$$

$$mL^2\frac{d^2\theta}{dt^2} = -(L)(mg\sin\theta) \tag{A.1}$$

Dividing both sides of the equation by mL^2 and transposing the torque to the left-side of the equation.

Answer 1: $d^2\theta/dt^2 + (g/L)\theta = 0$

For 2nd order ODEs, the coefficient of the zero-order derivative indicates the natural angular frequency. Recall,

$$\frac{d^2T}{dt^2} + c_p\frac{dT}{dt} + \omega_p^2 T = 0. \tag{A.2}$$

ω_p represents the natural angular frequency (recall the examples for mass–spring, RCL circuit, and rotating disks).

Comparing this equation with the 1st final answer,

$$\omega_p^2 = \frac{g}{L}$$

Answer 2: $\omega_p = \sqrt{g/L}$

2 Using the equation of ω_p derived in the 1st problem, the ratio of $\omega_{p_{\text{earth}}}$ and $\omega_{p_{\text{moon}}}$ can be represented by,

Solution:

Answer: Ratio $= \dfrac{\omega_{p_{\text{earth}}}}{\omega_{p_{\text{moon}}}} = \sqrt{\dfrac{g_{\text{earth}}}{g_{\text{moon}}}} \approx \sqrt{6}.$

3 Answer: The mass–spring system does not consider gravitational force

as influential to the internal/external system (i.e. negligible as an external force). On the other hand, the pendulum considers the influence of gravity as a significant internal parameter but not the mass.

4 Answer: We can start with eq. A.1 of the 1st problem and integrating it from an angular displacement θ_0 to θ_1. It is represented by the following:

$$\int_{\theta_0}^{\theta_1} mL^2 \frac{d^2\theta}{dt^2} d\theta = -\int_{\theta_0}^{\theta_1} mgL\sin\theta d\theta$$

Substituting $d\theta = (d\theta/dt)dt$, and at $t = 0$, $\theta = \theta_0$ and at $t = 1$, $\theta = \theta_1$,

$$\int_{t_0}^{t_1} mL^2 \frac{d^2\theta}{dt^2}\frac{d\theta}{dt} dt = -\int_{t_0}^{t_1} mgL\sin\theta \frac{d\theta}{dt} dt$$

Before moving forward, let us review the product rule:

$$\frac{d}{dt}\left(\frac{d\theta}{dt}\right)^2 = \frac{d}{dt}\left[\left(\frac{d\theta}{dt}\right)\left(\frac{d\theta}{dt}\right)\right] = 2\frac{d\theta}{dt}\frac{d^2\theta}{dt^2}$$

$$\frac{d\theta}{dt}\frac{d^2\theta}{dt^2} = \frac{1}{2}\frac{d}{dt}\left(\frac{d\theta}{dt}\right)^2$$

and the derivative of the cosine function, $d\cos\theta/dt = -\sin\theta d\theta/dt$
The equation above can be simplified further,

$$\int_{t_0}^{t_1} \frac{mL^2}{2}\frac{d}{dt}\left(\frac{d\theta}{dt}\right)^2 dt = \int_{t_0}^{t_1} mgL\frac{d\cos\theta}{dt} dt$$

From the relationship of linear velocity and angular velocity,

$$v = L\frac{d\theta}{dt},$$

we can integrate and substitute the angular velocity with linear velocity,

$$\frac{1}{2}mv_1^2 - \frac{1}{2}mv_0^2 = mg\cos\theta_1 - mg\cos\theta_0.$$

Rearranging the equation,

$$\frac{1}{2}mv_1^2 - mg\cos\theta_1 = \frac{1}{2}mv_0^2 - mg\cos\theta_0 = \text{constant}$$

Note: if the zero-reference is changed by a distance L upwards, potential energy mgL must be added to the equation.

5 We can start by the energy equation which is equal to the sum of kinetic and potential energy as follows,

$$\frac{1}{2}mv^2 + mgL\cos\theta = \text{constant}$$

Differentiating the equation with respect to t,

$$mv\frac{dv}{dt} - mgL\sin\theta\frac{d\theta}{dt} = 0$$

Recall the following,

$$v = L\frac{d\theta}{dt}$$

$$\frac{\mathrm{d}v}{\mathrm{d}\theta} = L\frac{\mathrm{d}^2\theta}{\mathrm{d}t^2}$$

The above equation is further simplified,

$$mL^2\left(\frac{\mathrm{d}\theta}{\mathrm{d}t}\right)\left(\frac{\mathrm{d}^2\theta}{\mathrm{d}t^2}\right) - mgL\sin\theta\frac{\mathrm{d}\theta}{\mathrm{d}t} = 0$$

Dividing both terms with $mL^2\left(\mathrm{d}\theta/\mathrm{d}t\right)$ results in,

$$\frac{\mathrm{d}^2\theta}{\mathrm{d}t^2} - \left(\frac{g}{L}\right)\sin\theta = 0.$$

If $\sin\theta \approx \theta$,

$$\frac{\mathrm{d}^2\theta}{\mathrm{d}t^2} - \left(\frac{g}{L}\right)\theta = 0.$$

4

4.1

1 Answer: $T = -16e^{-t} + 2e^{-2t} + 4t^2 - 12t + 14$

Homogeneous general solution: from $\lambda^2 + 3\lambda + 2 = 0, \lambda = -1,-2$

$T = C_1 e^{-t} + C_2 e^{-2t}$

Particular (special) solution: setting $T_s(t) = at^2 + bt + c$[185], $T_s' = 2at + b, T_s'' = 2a$

and substituting to the given ODE results in,

$(2a) + 3(2at + b) + 2(at^2 + bt + c) = 8t^2$. a,b,c can be solved,

$T_s(t) = 4t^2 - 12t + 14$[186]

General solution: $T = C_1 e^{-t} + C_2 e^{-2t} + 4t^2 - 12t + 14$

Given the initial conditions, C_1 and C_2 can be determined.

2 Answer: $T = -7 + 6e^t - t^2 - 2t$

Homogeneous general solution: from $\lambda^2 - \lambda = 0, \lambda = 0,1$

$T = C_1 + C_2 e^t$

Particular (special) solution: setting $T_s(t) = t \times (at + b) = at^2 + bt$[187], $T_s' = 2at + b, T_s'' = 2a$

and substituting to the given ODE results in,

$(2a) - (2at + b) = 2t$. a, b can be solved,

$T_s(t) = -t^2 - 2t$[188]

General solution: $T = C_1 + C_2 e^t - t^2 - 2t$

Given the initial conditions, C_1 and C_2 can be determined.

3 Answer: $T = -\cos t - 5\sin t + 4t$

Homogeneous general solution: from $\lambda^2 + 1 = 0, \lambda = \pm i$

$T = C_1 \cos t + C_2 \sin t$

Particular (special) solution: setting $T_s(t) = at + b$[189], $T_s' = a, T_s'' = 0$

and substituting to the given ODE results in,

$0 + at + b = 4t$. a, b can be solved,

$T_s(t) = 4t$[190]

General solution: $T = C_1 \cos t + C_2 \sin t + 4t$

Given the initial conditions, C_1 and C_2 can be determined.

4 Answer: $T = -e^t + 2e^{3t} - e^{2t}$

[185] Since the roots of the characteristic equation are real numbers and the external function is a quadratic equation, a particular solution can assume this form by utilizing group Ia in Table 4.1.

[186] $2at^2 + (6a + 2b)t + 2a + 3b + 2c = 8t^2$
$2a = 8$
$6a + 2b = 0$
$2a + 3b + 2c = 0$

[187] Since 0 is a root of the characteristic equation, group Ib in Table 4.1 is used.

[188] $-2a = 2 \quad 2a - b = 0$

[189] Since the root of the characteristic equation is imaginary which does not include 0, group Ia in Table 4.1 is used.

[190] $a = 4 \quad b = 0$

Homogeneous general solution: from $\lambda^2 - 4\lambda + 3 = 0, \lambda = 1,3$

$T = C_1 e^t + C_2 e^{3t}$

191 Since the root of the characteristic equation is not equal to the constant coefficient of the *exp* function of the external system, group IIa in Table 4.1 will be used.

192 $a = -1$

Particular (special) solution: setting $T_s(t) = ae^{2t}$ **◁191**, $T_s' = 2ae^{2t}$, $T_s'' = 4ae^{2t}$

and substituting to the given ODE results in,

$(4a - 8a + 3a)e^{2t} = e^{2t}$. a can be solved,

$T_s(t) = -e^{2t}$ **◁192**

General solution: $T = C_1 e^t + C_2 e^{3t} - e^{2t}$

Given the initial conditions, C_1 and C_2 can be determined.

5 Answer: $T = -\dfrac{7}{4}e^t + \dfrac{7}{4}e^{3t} - \dfrac{1}{2}te^t$

Homogeneous general solution: from $\lambda^2 - 4\lambda + 3 = 0, \lambda = 1,3$

$T = C_1 e^t + C_2 e^{3t}$

193 Since one of the roots of the characteristic equation is equal to the constant coefficient of the *exp* function of the external system, Group IIb in Table 4.1 applies.

194 $a = -1/2$

Particular (special) solution: setting $T_s(t) = ate^t$ **◁193**, $T_s' = ae^t + ate^t$, $T_s'' = 2ae^t + ate^t$

and substituting to the given ODE gives,

$[(2a + at) - 4(a + at) + 3at]e^t = e^t$. a can be solved,

$T_s(t) = -1/2te^t$ **◁194**

General solution: $T = C_1 e^t + C_2 e^{3t} - (1/2)te^t$

Given the initial conditions, C_1 and C_2 can be determined.

6 Answer: $T = 3te^{2t} + \dfrac{1}{2}t^2 e^{2t}$

Homogeneous general solution: from $\lambda^2 - 4\lambda + 4 = 0, \lambda = 2$

$T = (C_1 + C_2 t)e^{2t}$

195 Since the roots of the characteristic equation is a double root and is equal to the constant coefficient of the *exp* function of the external system, group IIc of Table 4.1 applies.

196 $a = 1/2$

Particular (special) solution: setting $T_s(t) = at^2 e^{2t}$ **◁195**, $T_s' = 2at^2 e^{2t} + 2ate^{2t}$,

$T_s'' = 4ate^{2t} + 4at^2 e^{2t} + 2ae^{2t} + 4ate^{2t}$,

$T_s'' = 4at^2 e^{2t} + 8ate^{2t} + 2ae^{2t}$

and substituting to the given ODE gives,

$[(4at^2 + 8at + 2a) - 4(2at^2 + 2at) + 4at^2]e^{2t} = e^{2t}$. a can be solved,

$T_s(t) = (1/2)t^2 e^{2t}$ **◁196**

General solution: $T = (C_1 + C_2 t)e^{2t} + (1/2)t^2 e^{2t}$

Given the initial conditions, C_1 and C_2 can be determined.

7 Answer: $T = -\dfrac{1}{8}\cos 3t + \sin 3t + \dfrac{1}{8}\cos t$

Homogeneous general solution: from $\lambda^2 + 9 = 0, \lambda = \pm 3i$

$T = C_1 \cos 3t + C_2 \sin 3t$

197 Since roots of the characteristic equation are $p \pm qi$ ($p = 0, q = 3$) and are neither equal to the constant coefficient of the cos function of the external system, group IIIa in Table 4.1 applies.

198 $a = 1/8, b = 0$

Particular (special) solution: setting $T_s(t) = a\cos t + b\sin t$ **◁197**,

$T_s' = -a\sin t + b\cos t$, $T_s'' = -a\cos t - b\sin t$

and substituting to the given ODE gives,

$-a\cos t - b\sin t + 9(a\cos t + b\sin t) = \cos t$. a,b can be solved,

$T_s(t) = (1/8)\cos t$ **◁198**

General solution: $T = C_1 \cos 3t + C_2 \sin 3t + (1/8)\cos t$

Given the initial conditions, C_1 and C_2 can be determined.

199 Since roots of the characteristic equation are $p \pm qi$ ($p = 0, q = 3$) and is equal to the constant coefficient of the cos function of the external system, group IIIb in Table 4.1 applies.

8 Answer: $T = \cos 3t + \sin 3t + t\sin 3t$

Homogeneous general solution: from $\lambda^2 + 9 = 0, \lambda = \pm 3i$

$T = C_1 \cos 3t + C_2 \sin 3t$

Particular (special) solution: setting $T_s(t) = t(a\cos 3t + b\sin 3t)$ **◁199**

$T'_s = a\cos 3t + b\sin 3t + t(-3a\sin 3t + 3b\cos 3t)$

$T''_s = -6a\sin 3t + 6b\cos 3t - 9t(a\cos 3t + b\sin 3t)$

$\quad = -6a\sin 3t + 6b\cos 3t - 9T_s$

and substituting to the given ODE gives,

$-6a\sin 3t + 6b\cos 3t - 9T_s + 9T_s = 6\cos 3t$. a, b can be solved,

$T_s(t) = t\sin 3t$ ↷**200**

General solution: $T = C_1\cos 3t + C_2\sin 3t + t\sin 3t$

Given the initial conditions, C_1 and C_2 can be determined.

↷**200** $a = 0, b = 1$

4.2

1 Multiply dX to the equation of motion and integrate from $X = X_0$ to $X = X_1$ ↷**201**,

$$\int_{X_0}^{X_1} \frac{d^2 X}{dt^2} dX + \int_{X_0}^{X_1} \omega_p^2 X dX = \int_{X_0}^{X_1} F_p\sin\omega t dX$$

Convert the integrator to the dependent variable t ↷**202**,

$$\int_{t_0}^{t_1}\left(\frac{d^2 X}{dt^2}\right)\left(\frac{dX}{dt}\right)dt + \int_{t_0}^{t_1}\omega_p^2 X\left(\frac{dX}{dt}\right)dt = \int_{t_0}^{t_1} F_p\sin\omega t\left(\frac{dX}{dt}\right)dt$$

$$\int_{t_0}^{t_1}\frac{1}{2}\frac{d}{dt}\left(\frac{dX}{dt}\right)^2 dt + \int_{t_0}^{t_1}\frac{1}{2}\omega_p^2\left(\frac{dX^2}{dt}\right)dt = \int_{t_0}^{t_1} F_p\sin\omega t\left(\frac{dX}{dt}\right)dt$$ ↷**203**

Noting that $v = dX/dt$, $v_0 = dX/dt|_{t=t_0}$, $v_1 = dX/dt|_{t=t_1}$

Answer: $\dfrac{1}{2}v_1^2 - \dfrac{1}{2}v_0^2 + \dfrac{1}{2}\omega_p^2 X_1^2 - \dfrac{1}{2}\omega_p^2 X_0^2 = \displaystyle\int_{t_0}^{t_1} F_p\sin\omega t \times v dt$ ↷**204**

2 Substituting the exact solution$[X = F_p/2\omega_p^2(\sin\omega_p t - t\omega_p\cos\omega_p t)]$ to the right-hand side of the above equation, $v = X' = (F_p/2)[t\sin(\omega_p t)]$

$$W = \int_0^{t_1} F_p\sin\omega t \times v dt = \int_0^{t_1}\frac{F_p^2}{2}t\sin\omega_p^2 t dt$$ ↷**205**

From the integrand $t\sin\omega_p^2 t \geq 0$ alone, it can be said that energy is bound to accumulate in the system as time t approaches infinity.

↷**201** As mentioned in ch. 3, if both side of the equation of motion are integrated from an initial displacement $X_0(t = 0)$ to a displacement of a certain time, $X_1(t = t_1)$, the amount of work exerted by the system in that range of displacement should obtained.

↷**202** $dX = (dX/dt)dt$

↷**203** $(d/dt)(dX/dt)^2 = 2(dX/dt)$ $(d^2 X/dt^2)$, $dX^2/dt = 2X(dX/dt)$

↷**204** The left-hand side terms is the amount of change in kinetic energy and spring's elastic potential energy from t_0 to t_1. The right-hand side is the work done by the external force from t_0 to t_1 = energy stored in the system.

↷**205** $\omega = \omega_p$

4.3
If the external force is replaced with $g(t) = F_0 e^{-2t}$, the ODE of the mass–spring system will become, $d^2 X/dt^2 + \omega_p^2 X = F_p e^{-2t}$
The general solution of the homogeneous form of the above non-homogeneous ODE is $X = C_1\cos\omega_p t + C_2\sin\omega_p t$. The non-homogeneous particular solution is $X_s = ae^{-2t}$ ↷**206** ↷**207**. Substituting X_s to the above non-homogeneous ODE,

$4ae^{-2t} + a\omega_p^2 e^{-2t} = F_p e^{-2t}$ appears. Hence, a is as follows,

$$a = \frac{F_p}{4 + \omega_p^2}$$

The general solution of the non-homogeneous IDE is

$$X = C_1\cos\omega_p t + C_2\sin\omega_p t + \frac{F_p e^{-2t}}{4 + \omega_p^2}.$$

The general solution of the homogeneous ODE corresponds to a natural

↷**206** $X'_s = -2ae^{-2t}$, $X''_s = 4ae^{-2t}$.

↷**207** $\lambda = \pm\omega_p i$ is an imaginary number which does not equate to the constant coefficient (-2) of the exp function regardless of the value of ω_p.

⌐**208** $C_1 \cos \omega_p t + C_2 \sin \omega_p t$
$= \sqrt{C_1^2 + C_2^2} \cos (\omega_p t - \eta)$

vibration⌐**208**where as that of the particular (special) solution suggests attenuation with time. Therefore, the system initially vibrates naturally, and then its vibration is attenuated with time.

5

5.1   Replace the 1st derivative of T with $g(t,T)$ in the Taylor-series expansion.

⌐**209** $O(\Delta t^3)$ is achieved by rounding off terms with Δt powers higher than 3.

⌐**210** The approximation of 2nd order precision is used which leaves only terms with powers of Δt no greater than 2.

$$T(t + \Delta t) = T(t) + \frac{\Delta t}{1!}\frac{dT(t)}{dt} + \frac{\Delta t^2}{2!}\frac{d^2T(t)}{dt^2} + O(\Delta t^3) \quad ^{⌐\mathbf{209}} \qquad (6.9\text{a})$$

$$T(t + \Delta t) \cong T(t) + \frac{\Delta t}{1!}g(t,T) + \frac{\Delta t^2}{2!}\frac{dg(t,T)}{dt} \quad ^{⌐\mathbf{210}} \qquad (6.9\text{b})$$

Note that g is a function of both t and T such that,

$$\frac{dg(t,T)}{dt} = \frac{\partial g}{\partial t}\frac{dt}{dt} + \frac{\partial g}{\partial T}\frac{dT}{dt} = \frac{\partial g}{\partial t} + \frac{\partial g}{\partial T} \times g \qquad (6.10)$$

Substituting eq. 6.10 to eq. 6.9b yields a Taylor-series expansion that is of 2nd-order accuracy. The following equation is derived after further substituting $T(t + \Delta t)$, $T(t)$, and g with $T_{n+1}, T_n, g(t_n, T_n)$, respectively.

$$T_{n+1} \cong T_n + \Delta t\, g(t_n, T_n) + \frac{\Delta t^2}{2}\left(\frac{\partial g}{\partial t} + \frac{\partial g}{\partial T}g(t_n, T_n)\right) \qquad (6.11)$$

Now, let's decide the weighting coefficient that satisfies eq. 6.11 by Taylor's series expansion of the provisional prediction values of the two projections. The second temporary prediction value is Taylor-expanded around the first prediction value. In other words, Taylor's series expansion of $T_{n+1} = T_n + (\alpha h_1 + \beta h_2)$ around $g(t_n, T_n)$ leads to,

$$g(t_n + \Delta t, T_n + h_1) = g(t_n, T_n) + \Delta t\left(\frac{\partial g}{\partial t}\right) + h_1\left(\frac{\partial g}{\partial T}\right) \qquad (6.12)$$

$$= g(t_n, T_n) + \Delta t\left(\frac{\partial g}{\partial t}\right) + \Delta t\left(\frac{\partial g}{\partial T}\right)g(t_n, T_n) \qquad (6.13)$$

Thus,

$$h_2 = \Delta t\, g(t_n, T_n) + \Delta t^2\left(\frac{\partial g}{\partial t} + \frac{\partial g}{\partial T}g(t_n, T_n)\right) \qquad (6.14)$$

$$h_1 = \Delta t\, g(t_n, T_n) \qquad (6.15)$$

For $T_{n+1} = T_n + (\alpha h + \beta h_2)$ to satisfy eq. 6.11, α and β must be equated to half of h_1 and h_2, interchangeably.

$$T_{n+1} = T_n + \frac{1}{2}(h_1 + h_2) \qquad (6.16)$$

This algorithm is sometimes referred to as improved Euler method, Heun method, or 2nd order Runge–Kutta method.

5.2   The following script may be used to compare the exact solution with numerical methods, Euler and 4th order Runge–Kutta.

```python
1   import numpy as np
2   import matplotlib.pyplot as plt
3
4   def exact(t):
5           return 2*np.exp(-5.*t)
6
7   def euler(t,dt):
8           T = np.copy(t)
9           T[0] = 2.
10          for i in range(t.shape[0]-1):
11                  h1 = dt*g(T[i])
12                  T[i+1] = T[i]+h1
13          return T
14
15  def rk(t,dt):
16          T = np.copy(t)
17          T[0] = 2.
18          for i in range(t.shape[0]-1):
19                  h1 = dt*g(T[i])
20                  h2 = dt*g(T[i]+0.5*h1)
21                  h3 = dt*g(T[i]+0.5*h2)
22                  h4 = dt*g(T[i]+h3)
23                  T[i+1] = T[i] + (h1+\
24                                  2.*h2+\
25                                  2.*h3+\
26                                  h4)*(1/6.)
27          return T
28
29  def g(T):
30          return -5.*T
31
32  #Initialize values.
33  dt = 0.1
34  t = np.arange(0.,1.+dt,dt)
35
36  #Solve
37  T_exact = exact(t)
38  T_euler = euler(t,dt)
39  T_rk    = rk(t,dt)
40
41  #Plot results
42  plt.plot(t,T_exact,color='k',label='Exact')
43  plt.plot(t,T_euler,color='k',linestyle='dashed'\
44          ,label='Euler')
45  plt.plot(t,T_rk,marker='o',color='k',linestyle='
    None'\
46          ,label='4th Order RK')
47  plt.legend()
48  plt.xlabel(r'$t$')
49  plt.ylabel(r'$T$')
50  plt.show()
```

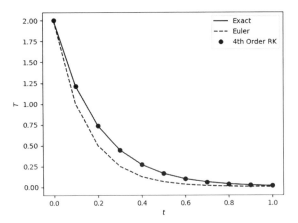

Figure 1 Difference in solution methods for $T' + 5T = 0$ with $T(0) = 3$

6

6.1 Lorenz attractor

The following code may be used to solve for X, Y, and Z according to the given specifications.

```
1   import numpy as np
2   import matplotlib.pyplot as plt
3
4   def rk(t,dt,xinit,yinit,zinit):
5       X = np.copy(t);Y = np.copy(t);Z = np.copy(
             t)
6       X[:] = 0.; Y[:] = 0.; Z[:] = 0.
7       X[0] = xinit; Y[0] = yinit; Z[0] = zinit
8       for i in range(0,t.shape[0]-1):
9           h1 = dt*dXdt(X[i],Y[i],Z[i])
10          j1 = dt*dYdt(X[i],Y[i],Z[i])
11          k1 = dt*dZdt(X[i],Y[i],Z[i])
12          h2 = dt*dXdt(X[i]+0.5*h1,Y[i]+0.5*
                 j1,Z[i]+0.5*k1)
13          j2 = dt*dYdt(X[i]+0.5*h1,Y[i]+0.5*
                 j1,Z[i]+0.5*k1)
14          k2 = dt*dZdt(X[i]+0.5*h1,Y[i]+0.5*
                 j1,Z[i]+0.5*k1)
15          h3 = dt*dXdt(X[i]+0.5*h2,Y[i]+0.5*
                 j2,Z[i]+0.5*k2)
16          j3 = dt*dYdt(X[i]+0.5*h2,Y[i]+0.5*
                 j2,Z[i]+0.5*k2)
17          k3 = dt*dZdt(X[i]+0.5*h2,Y[i]+0.5*
                 j2,Z[i]+0.5*k2)
18          h4 = dt*dXdt(X[i]+h3,Y[i]+j3,Z[i]+
                 k3)
19          j4 = dt*dYdt(X[i]+h3,Y[i]+j3,Z[i]+
                 k3)
20          k4 = dt*dZdt(X[i]+h3,Y[i]+j3,Z[i]+
                 k3)
21          X[i+1] = X[i]+(h1+2.*h2+2.*h3+h4)
                 /6.
22          Y[i+1] = Y[i]+(j1+2.*j2+2.*j3+j4)
                 /6.
```

```
23                       Z[i+1] = Z[i]+(k1+2.*k2+2.*k3+k4)
                         /6.
24            return X,Y,Z
25
26  def dXdt(X,Y,Z):
27            return -sigma*(X-Y)
28
29  def dYdt(X,Y,Z):
30            return -X*Z+gamma*X-Y
31
32  def dZdt(X,Y,Z):
33            return X*Y-b*Z
```

To construct the time-series diagram for further analyses, append the codes after the functions above to create the following figure.

```
1   '''
2   To␣omit␣the␣necessity␣of␣passing␣sigma,gamma,b␣as␣
        arguments,
3   they␣can␣be␣set␣as␣global␣variables␣recognized␣
        throughout.
4   '''
5   global sigma,gamma,b
6   sigma = 10.
7   gamma = 28.
8   b = 8./3.
9
10  dt = 0.01
11  t = np.arange(0.,30.+dt,dt)
12  X1,Y1,Z1 = rk(t,dt,1.,2.,10.)
13  X2,Y2,Z2 = rk(t,dt,1.1,2.,10.)
14
15  plt.plot(t,X1,color='k',linestyle='-',label=r'case
        ␣1␣$X$')
16  plt.plot(t,Z1,color='k',linestyle='--',label=r'
        case␣1␣$Z$')
17  plt.plot(t,X2,color='grey',linestyle='-',label=r'
        case␣2␣$X$')
18  plt.plot(t,Z2,color='grey',linestyle='--',label=r'
        case␣2␣$Z$')
19  plt.legend(loc='upper␣center', bbox_to_anchor
        =(0.5, 1.1),
20               ncol=4, fancybox=True, shadow=True)
21  plt.xlabel(r'$t$')
22  plt.ylabel(r'$X,Z$')
23  plt.show()
```

The above script will create the following time-series plot of the *X* and *Z* values of each case.

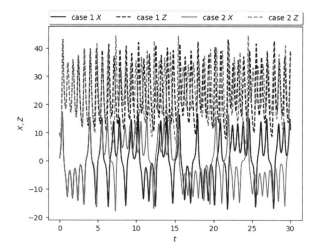

Figure 1 Time-series plot

To construct the X-Y phase diagram, the following lines of code for plotting may be used instead of the previous lines for plotting.

```
1  plt.plot(X1,Y1,color='k',linestyle='-',label=r'
     case 1')
2  plt.plot(X2,Y2,color='grey',linestyle='-',label=r'
     case 2')
3  plt.xlabel(r'$X$')
4  plt.ylabel(r'$Y$')
5  plt.legend()
6  plt.show()
```

The X-Y phase diagram will be produced.

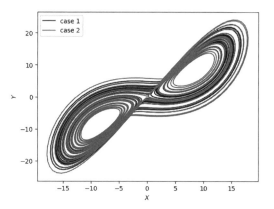

Figure 2 X-Y phase diagram

To construct the Y-Z phase diagram, the following lines of code for plotting may be used instead of the previous lines for plotting.

```
1  plt.plot(Y1,Z1,color='k',linestyle='-',label=r'
      case␣1')
2  plt.plot(Y2,Z2,color='grey',linestyle='-',label=r'
      case␣2')
3  plt.xlabel(r'$Y$')
4  plt.ylabel(r'$Z$')
5  plt.legend()
6  plt.show()
```

The Y-Z phase diagram will be produced.

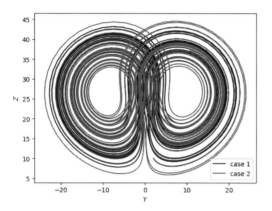

Figure 3 Y-Z phase diagram

6.2 Rössler attractor

Using the 4th order Runge–Kutta scheme, the following script may be used to set-up the system represented by the non-linear ODEs.

```
1  import numpy as np
2  import matplotlib.pyplot as plt
3
4  def rk(t,dt,xinit,yinit,zinit):
5        X = np.copy(t);Y = np.copy(t);Z = np.copy(
          t)
6        X[:] = 0.; Y[:] = 0.; Z[:] = 0.
7        X[0] = xinit; Y[0] = yinit; Z[0] = zinit
8        for i in range(0,t.shape[0]-1):
9              h1 = dt*dXdt(X[i],Y[i],Z[i])
10             j1 = dt*dYdt(X[i],Y[i],Z[i])
11             k1 = dt*dZdt(X[i],Y[i],Z[i])
12             h2 = dt*dXdt(X[i]+0.5*h1,Y[i]+0.5*
                 j1,Z[i]+0.5*k1)
13             j2 = dt*dYdt(X[i]+0.5*h1,Y[i]+0.5*
                 j1,Z[i]+0.5*k1)
14             k2 = dt*dZdt(X[i]+0.5*h1,Y[i]+0.5*
                 j1,Z[i]+0.5*k1)
15             h3 = dt*dXdt(X[i]+0.5*h2,Y[i]+0.5*
                 j2,Z[i]+0.5*k2)
16             j3 = dt*dYdt(X[i]+0.5*h2,Y[i]+0.5*
                 j2,Z[i]+0.5*k2)
17             k3 = dt*dZdt(X[i]+0.5*h2,Y[i]+0.5*
                 j2,Z[i]+0.5*k2)
```

```
18                      h4 = dt*dXdt(X[i]+h3,Y[i]+j3,Z[i]+
                          k3)
19                      j4 = dt*dYdt(X[i]+h3,Y[i]+j3,Z[i]+
                          k3)
20                      k4 = dt*dZdt(X[i]+h3,Y[i]+j3,Z[i]+
                          k3)
21                      X[i+1] = X[i]+(h1+2.*h2+2.*h3+h4)
                          /6.
22                      Y[i+1] = Y[i]+(j1+2.*j2+2.*j3+j4)
                          /6.
23                      Z[i+1] = Z[i]+(k1+2.*k2+2.*k3+k4)
                          /6.
24              return X,Y,Z
25
26      def dXdt(X,Y,Z):
27              return -Y-Z
28
29      def dYdt(X,Y,Z):
30              return X+a*Y
31
32      def dZdt(X,Y,Z):
33              return b+Z*(X-c)
```

Using the functions created above, the required analyses can be conducted. The scripts are quite similar to the previous exercise; except this time, there will be 3 cases where each case has a corresponding c value. The script for constructing the time-series of X and Z are as follows.

```
1   '''
2   To␣omit␣the␣necessity␣of␣passing␣a,␣b,␣c␣as␣
      arguments,
3   they␣can␣be␣set␣as␣global␣variables␣recognized␣
      throughout.
4   '''
5   global a,b,c
6   a = 0.1
7   b = 0.1
8   dt = 0.01
9   t = np.arange(0.,200.+dt,dt)
10
11  c=4.
12  X1,Y1,Z1 = rk(t,dt,1.,1.,1.)
13  c=12.
14  X2,Y2,Z2 = rk(t,dt,1.,1.,1.)
15  c=8.
16  X3,Y3,Z3 = rk(t,dt,1.,1.,1.)
17
18  plt.plot(t,X1,color='k',linestyle='-',label=r'case
      ␣1␣$X$')
19  plt.plot(t,Z1,color='k',linestyle='--',label=r'
      case␣1␣$Z$')
20  plt.plot(t,X2,color='grey',linestyle='-',label=r'
      case␣2␣$X$')
21  plt.plot(t,Z2,color='grey',linestyle='--',label=r'
      case␣2␣$Z$')
22  plt.plot(t,X3,color=(.35,.35,.35),linestyle='-',
      label=r'case␣2␣$X$')
23  plt.plot(t,Z3,color=(.35,.35,.35),linestyle='--',
      label=r'case␣2␣$Z$')
```

```
24
25   plt.legend(loc='upper␣center', bbox_to_anchor
       =(0.5, 1.15),
26              ncol=3, fancybox=False, shadow=False)
27   plt.xlabel(r'$t$')
28   plt.ylabel(r'$X,Z$')
29   plt.show()
```

This will produce a time-series plot with X and Z plotted together for all cases.

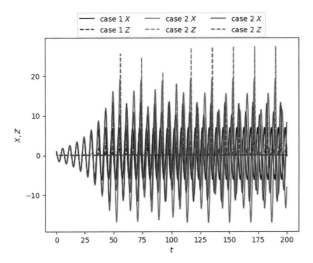

Figure 4   Time-series plot

The script for constructing the X-Y phase diagram is as follows.

```
1   plt.plot(X1,Y1,color='k',linestyle='-',label=r'
      case␣1')
2   plt.plot(X2,Y2,color='grey',linestyle='-',label=r'
      case␣2')
3   plt.plot(X3,Y3,color=(.35,.35,.35),linestyle='-',
      label=r'case␣3')
4   plt.xlabel(r'$X$')
5   plt.ylabel(r'$Y$')
6   plt.legend()
7   plt.show()
```

The script above will produce the X-Y phase diagram with all cases included.

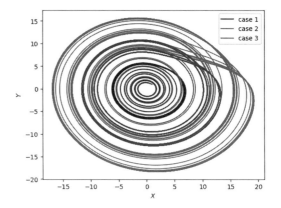

Figure 5 *X-Y* phase diagram

The script for constructing the *Y-Z* phase diagram is as follows.

```
1  plt.plot(Y1,Z1,color='k',linestyle='-',label=r'
     case␣1')
2  plt.plot(Y2,Z2,color='grey',linestyle='-',label=r'
     case␣2')
3  plt.plot(Y3,Z3,color=(.35,.35,.35),linestyle='-',
     label=r'case␣3')
4  plt.xlabel(r'$Y$')
5  plt.ylabel(r'$Z$')
6  plt.legend()
7  plt.show()
```

The script above will produce the *Y-Z* phase diagram with all cases included.

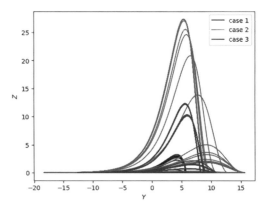

Figure 6 *Y-Z* phase diagram

A List of formulae

A.1 Trigonometric identites

- Pythagorean Identities

$$\sin^2(t) + \cos^2(t) = 1$$

- Quotient Identities

$$\tan(t) = \frac{\sin(t)}{\cos(t)} \quad \cot(t) = \frac{\cos(t)}{\sin(t)}$$

- Co-Function Identities

$$\sin\left(\frac{\pi}{2} - t\right) = \cos(t) \quad \cos\left(\frac{\pi}{2} - t\right) = \sin(t)$$

- Parity Identities (Even & Odd)

$$\sin(-t) = -\sin(t) \quad \cos(-t) = \cos(t)$$

- Sum & Difference Formulae

$$\sin(t_1 \pm t_2) = \sin(t_1)\cos(t_2) \pm \cos(t_1)\sin(t_2)$$
$$\cos(t_1 \pm t_2) = \cos(t_1)\cos(t_2) \mp \sin(t_1)\sin(t_2)$$
$$\tan(t_1 \pm t_2) = \frac{\tan(t_1) \pm \tan(t_2)}{1 \mp \tan(t_1)\tan(t_2)}$$

- Double Angle Formulae

$$\sin(2t) = 2\sin(t)\cos(t)$$
$$\cos(2t) = \cos^2(t) - \sin^2(t)$$
$$\cos(2t) = 2\cos^2(t) - 1$$
$$\cos(2t) = 1 - 2\sin^2(t)$$
$$\tan(2t) = \frac{2\tan(t)}{1 - \tan^2(t)}$$

- Power-Reducing/Half Angle Formulae

$$\sin^2(t) = \frac{1 - \cos(2t)}{2}$$
$$\cos^2(t) = \frac{1 + \cos(2t)}{2}$$
$$\tan^2(t) = \frac{1 - \cos(2t)}{1 + \cos(2t)}$$

- Sum-to-Product Formulae

$$\sin(t_1) + \sin(t_2) = 2\sin\left(\frac{t_1 + t_2}{2}\right)\cos\left(\frac{t_1 - t_2}{2}\right)$$
$$\sin(t_1) - \sin(t_2) = 2\cos\left(\frac{t_1 + t_2}{2}\right)\sin\left(\frac{t_1 - t_2}{2}\right)$$
$$\cos(t_1) + \cos(t_2) = 2\cos\left(\frac{t_1 + t_2}{2}\right)\cos\left(\frac{t_1 - t_2}{2}\right)$$
$$\cos(t_1) - \cos(t_2) = -2\sin\left(\frac{t_1 + t_2}{2}\right)\sin\left(\frac{t_1 - t_2}{2}\right)$$

- Product-to-Sum Formulae

$$\sin(t_1)\sin(t_2) = \frac{1}{2}[\cos(t_1 - t_2) - \cos(t_1 + t_2)]$$
$$\cos(t_1)\cos(t_2) = \frac{1}{2}[\cos(t_1 - t_2) + \cos(t_1 + t_2)]$$
$$\sin(t_1)\cos(t_2) = \frac{1}{2}[\sin(t_1 + t_2) + \sin(t_1 - t_2)]$$
$$\cos(t_1)\sin(t_2) = \frac{1}{2}[\sin(t_1 + t_2) - \sin(t_1 - t_2)]$$

A.2   Differentiation and integration formulae

Differentiation.

$(ct)' = ct'$ (c constant)

$(t_1 + t_2)' = t_1' + t_2'$

$(t_1 t_2)' = t_1' t_2 + t_1 t_2'$

$\left(\dfrac{t_1}{t_2}\right)' = \dfrac{t_1' t_2 - t_1 t_2'}{t_2^2}$

$\dfrac{\mathrm{d}t}{\mathrm{d}x} = \dfrac{\mathrm{d}t}{\mathrm{d}y} \cdot \dfrac{\mathrm{d}y}{\mathrm{d}x}$ (chain rule)

$(x^n)' = n x^{n-1}$

$(e^x)' = e^x$

$(e^{ax})' = a e^{ax}$

$(a^x)' = a^x \ln a$

$(\sin x)' = \cos x$

$(\cos x)' = -\sin x$

$(\ln x)' = \dfrac{1}{x}$

Integration.

$\displaystyle\int t_1 t_2' \mathrm{d}x = t_1 t_2 - \int t_1' t_2 \mathrm{d}x$ (by parts)

$\displaystyle\int x^n \mathrm{d}x = \dfrac{x^{n+1}}{n+1} + c$ ($n \neq -1$)

$\displaystyle\int \dfrac{1}{x} \mathrm{d}x = \ln(|x|) + c$

$\displaystyle\int e^{ax} \mathrm{d}x = \dfrac{1}{a} e^{ax} + c$

$\displaystyle\int \sin x \mathrm{d}x = -\cos x + c$

$\displaystyle\int \cos x \mathrm{d}x = \sin x + c$

Index

B

beats phenomenon, 65, 67
Beer's law, 29
boundary condition, 7, 17
box model, 30

C

capacitor, 27, 69
chaos, 104
chaos phenomenon, 103
characteristic equation, 40
control volume, 31

D

derivatives, 6
differential equations, 6
Dirichlet condition, 18
discretization, 87
disk rotation, 49
distinct real root, 41

E

energy conservation, 50
equation of motion, 51
Euler method, 86
external system, 13

F

feedback, 24
    negative, 24
    positive, 24
flux, 31
free-falling body, 25

H

half-life, 29
harmonic oscillation, 45, 66

I

imaginary root, 42
inductor, 69
initial condition, 7, 17
internal system, 13

K

Kirchhoff's voltage law, 28, 48

L

law, 15
Logistic model, 32
Lorenz model, 119
Lotka–Volterra equation, 111

M

mass–spring system, 43
model, 15
momentum conservation, 50

N

Neumann condition, 18
numerics, 85

O

Ohm's law, 28
order, 7
ordinary differential equations, 6
    homogeneous, 9
    linear, 8
    non-linear, 8
    separable, 23
    simultaneous, 7, 104

P

partial differential equations, 6
phase diagram, 13

R

radiation, 29
real double root, 41
resistance coefficient, 45
resistor, 27, 69
resolution, 86
resonance, 60, 71
    practical, 71
RLC electric circuit, 68
Runge–Kutta method, 89

S

solution, 4
    approximate, 4
    basis of, 40
    empirical, 5
    exact, 4, 8
    general, 8
    particular, 8
spring constant, 45

strange attractors, 105
superposition principle, 11, 39
system
        critically damped, 45, 46
        damped, 45
        overdamped, 47
        undamped, 45
        underdamped, 45, 46

T

Taylor expansion, 86
time-series diagram, 13
tortion, 49
turbidity, 29

V

variable separation, 23
variables, 5
        dependent, 5
        independent, 5

編著者略歴

神 田　　学　　Manabu Kanda
1964年　新潟県に生まれる
1988年　東京工業大学大学院理工学研究科
　　　　土木工学専修修士課程修了
現　在　東京工業大学環境・社会理工学院教授
　　　　博士（工学）

訳者略歴

アルビン・バルケズ　　Alvin C. G. Varquez
1987年　フィリピン・ボホール州に生まれる
2014年　東京工業大学大学院理工学研究科
　　　　国際開発工学専攻博士課程修了
現　在　東京工業大学環境・社会理工学院准教授
　　　　博士（工学）

Ordinary Differential Equations
and Physical Phenomena
— A Short Introduction with Python —　　　　　定価はカバーに表示

2020 年 11 月 1 日　初版第 1 刷

著　者　　神 田　　学

訳　者　　アルビン・バルケズ

発行者　　朝 倉 誠 造

発行所　　株式会社 朝 倉 書 店

東京都新宿区新小川町 6-29
郵 便 番 号　　162-8707
電　話　03（3260）0141
Ｆ Ａ Ｘ　03（3260）0180
http://www.asakura.co.jp

〈検印省略〉

ⓒ 2020 〈無断複写・転載を禁ず〉　　　　シナノ印刷・渡辺製本

ISBN 978-4-254-20169-7　C 3050　　　　Printed in Japan

山口大 岡本和也・山口大 福代和宏著　山口大 上西　研監修

MOT 研究開発マネジメント入門

技術経営（MOT）の標準的な思考技術を習得する教科書。コア技術の理解・市場の分析・最適な組織の構築などの視点から研究開発マネジメントを理解。〔内容〕研究開発の流れ／提案と実践／組織構築と人財育成／長期ビジョン／他

20167-3 C3050　　　A 5 判 232頁 本体3500円

Peirce, J. 他著　京大 蘆田　宏・愛媛大 十河宏行監訳

PsychoPyでつくる心理学実験

心理学実験作成環境の開発者による解説書。プログラミングなしに作成可能な基本から，Pythonによる上級者向けの調整まで具体的な事例を通して解説。〔内容〕画像／タイミング・刺激／フィードバック／無作為化／アイトラッキング／他

52029-3 C3011　　　A 5 判 328頁 本体4800円

Yuxi (Hayden) Liu著　黒川利明訳

事例とベストプラクティス Python 機械学習
基本実装とscikit-learn/TensorFlow/PySpark活用

人工知能のための機械学習の基本，重要なアルゴリズムと技法，実用的なベストプラクティス。【例】テキストマイニング，教師あり学習によるオンライン広告クリックスルー予測，学習のスケールアップ（Spark），回帰による株価予測

12244-2 C3041　　　A 5 判 304頁 本体3900円

Theodore Petrou著　黒川利明訳

pandas クックブック
—Pythonによるデータ処理のレシピ—

データサイエンスや科学計算に必須のツールを詳説。〔内容〕基礎／必須演算／データ分析開始／部分抽出／booleanインデックス法／インデックスアライメント／集約，フィルタ，変換／整然形式／オブジェクトの結合／時系列分析／可視化

12242-8 C3004　　　A 5 判 384頁 本体4200円

慶大 中妻照雄著
実践Pythonライブラリー

Pythonによる ベイズ統計学入門

ベイズ統計学を基礎から解説，Pythonで実装。マルコフ連鎖モンテカルロ法にはPyMC3を活用。〔内容〕「データの時代」におけるベイズ統計学／ベイズ統計学の基本原理／様々な確率分布／PyMC／時系列データ／マルコフ連鎖モンテカルロ法

12898-7 C3341　　　A 5 判 224頁 本体3400円

愛媛大 十河宏行著
実践Pythonライブラリー

はじめてのPython & seaborn
—グラフ作成プログラミング—

作図しながらPythonを学ぶ〔内容〕準備／いきなり棒グラフを描く／データの表現／ファイルの読み込み／ヘルプ／いろいろなグラフ／日本語表示と制御文／ファイルの実行／体裁の調整／複合的なグラフ／ファイルへの保存／データ抽出と関数

12897-0 C3341　　　A 5 判 192頁 本体3000円

海洋大 久保幹雄監修　小樽商大 原口和也著
実践Pythonライブラリー

Kivy プログラミング
—Pythonでつくるマルチタッチアプリ—

スマートフォンで使えるマルチタッチアプリをPython Kivyで開発。〔内容〕ウィジェットとプロパティ／イベントとプロパティ／KV言語／キャンバス／サンプルアプリの開発／次のステップに向けて／ウィジェット・リファレンス／他

12896-3 C3341　　　A 5 判 200頁 本体3200円

前東大 小柳義夫訳
実践Pythonライブラリー

計 算 物 理 学 I
—数値計算の基礎/HPC/フーリエ・ウェーブレット解析—

Landau et al., Computational Physics: Problem Solving with Python, 3rd ed. を2分冊で。理論からPythonによる実装まで解説。〔内容〕誤差／モンテカルロ法／微積分／行列／データのあてはめ／微分方程式／HPC／フーリエ解析／他

12892-5 C3341　　　A 5 判 376頁 本体5400円

前東大 小柳義夫監訳
実践Pythonライブラリー

計 算 物 理 学 II
—物理現象の解析・シミュレーション—

計算科学の基礎を解説したI巻につづき，II巻ではさまざまな物理現象を解析・シミュレーションする。〔内容〕非線形系のダイナミクス／フラクタル／熱力学／分子動力学／静電場解析／熱伝導／波動方程式／衝撃波／流体力学／量子力学／他

12893-2 C3341　　　A 5 判 304頁 本体4600円

海洋大 久保幹雄監修　東邦大 並木　誠著
実践Pythonライブラリー

Pythonによる 数理最適化入門

数理最適化の基本的な手法をPythonで実践しながら身に着ける。初学者にも試せるようにプログラミングの基礎から解説。〔内容〕Python概要／線形最適化／整数線形最適化問題／グラフ最適化／非線形最適化／付録:問題の難しさと計算量

12895-6 C3341　　　A 5 判 208頁 本体3200円

慶大 中妻照雄著
実践Pythonライブラリー

Pythonによる ファイナンス入門

初学者向けにファイナンスの基本事項を確実に押さえた上で，Pythonによる実装をプログラミングの基礎から丁寧に解説。〔内容〕金利・現在価値・内部収益率・債権分析／ポートフォリオ選択／資産運用における最適化問題／オプション価格

12894-9 C3341　　　A 5 判 176頁 本体2800円

愛媛大 十河宏行著
実践Pythonライブラリー

心理学実験プログラミング
—Python/PsychoPyによる実験作成・データ処理—

Python（PsychoPy）で心理学実験の作成やデータ処理を実践。コツやノウハウも紹介。〔内容〕準備（プログラミングの基礎など）／実験の作成（刺激の作成，計測）／データ処理（整理，音声，画像）／付録（セットアップ，機器制御）

12891-8 C3341　　　A 5 判 192頁 本体3000円

東工大 花岡伸也編著
シリーズ〈新しい工学〉2

プロジェクトマネジメント入門

20522-0　C3350　　　　　Ｂ５判 144頁 本体2800円

工学の視点からプロジェクトマネジメントの基礎理論と実践・ケーススタディまでをコンパクトに解説。太陽光発電や国際開発プロジェクトなど多くの事例を収録し，グローバル時代の実務家の基礎教養となるべき知識を提供する。

前東工大 大即信明・東工大 日野出洋文・前東工大 サリム・クリス著
シリーズ〈新しい工学〉3

材　料　科　学

20523-7　C3350　　　　　Ｂ５判 148頁 本体2800円

機械系，電子系，建設系など多岐にわたる現代の材料工学の共通の基礎を学べる入門書。〔内容〕原子構造と結合／結晶構造／固体の不完全性／拡散／状態図／電気的性質／電気化学的性質／光学的性質および超伝導材料／磁気的性質

前東工大 大即信明・東工大 中崎清彦編著
シリーズ〈新しい工学〉4

工　業　材　料
—エンジニアリングからバイオテクノロジーまで—

20524-4　C3350　　　　　Ｂ５判 152頁 本体2800円

無機・金属材料から，高分子材料・生物材料まで，幅広いトピックをバランス良く記述した教科書。現代的な材料工学の各分野を一望できるよう，基礎から先端までの具体的な例を多数取り上げ，幅広い知識をやさしく解説した。

東工大 山下幸彦著
シリーズ〈新しい工学〉5

線　形　シ　ス　テ　ム　論

20525-1　C3350　　　　　Ｂ５判 152頁 本体2800円

回路解析，制御回路，質点系の力学の解析の基礎ツールである線形システム論を丁寧に解説する。三角関数・指数関数の復習から，フーリエ変換，ラプラス変換，連続時間線形システム，離散時間信号の変換，離散時間線形システムを学ぶ。

明大中野 西村重人訳　前九大 高瀬正仁監訳

フーリエ　熱の解析的理論

11156-9　C3041　　　　　Ａ５判 500頁 本体10000円

後世の数学・科学技術に多大な影響を与えた19世紀フランスの数学者フーリエの主著の全訳。熱伝播の問題のためにフーリエが編み出した数学と，彼の自然思想が展開される。学術的知見に基づいた正確な翻訳に，豊富な注釈・解説を付す。

前東工大 日野幹雄著

乱　流　の　科　学
—構造と制御—

20161-1　C3050　　　　　Ａ５判 1152頁 本体26000円

乱流の作用は運動量・物質・熱などの拡散，混合とエネルギーの消散である。乱流なくして我々は呼吸もできない。乱流状態がいかにして発生し作用するか，その基礎的メカニズムを詳細にかつできるだけ数式を使わずに解説した珠玉の解説書。

鳥取大 田村篤敬・岡山大 柳瀬眞一郎・岡山大 河内俊憲著

工 学 の た め の 物 理 数 学

20168-0　C3050　　　　　Ａ５判 200頁 本体3200円

工学部生が学ぶ応用数学の中でも，とくに「これだけは知っていたい」というテーマを3章構成で集約。例題や練習問題を豊富に掲載し，独習にも適したテキストとなっている。〔内容〕複素解析／フーリエ・ラプラス解析／ベクトル解析。

東北大 高　偉・東北大 清水裕樹・東北大 羽根一博・東北大 祖山　均・東北大 足立幸志著
Bilingual edition

計測工学 Measurement and Instrumentation

20165-9　C3050　　　　　Ａ５判 200頁 本体2800円

計測工学の基礎を日本語と英語で記述。〔内容〕計測の概念／計測システムの構成と特性／計測の不確かさ／信号の変換／データ処理／変位と変形／速度と加速度／力とトルク／材料物性値／流体／温度と湿度／光／電気磁気／計測回路

前お茶女大 河村哲也監訳　前お茶女大 井元　薫訳

高 等 数 学 公 式 便 覧

11138-5　C3342　　　　　菊判 248頁 本体4800円

各公式が，独立にページ毎の囲み枠によって視覚的にわかりやすく示され，略図も多用しながら明快に表現され，必要に応じて公式の使用法を例を用いながら解説。表・裏扉に重要な公式を掲載，豊富な索引付き。〔内容〕数と式の計算／幾何学／初等関数／ベクトルの計算／行列，行列式，固有値／数列，級数／微分法／積分法／微分幾何学／各変数の関数／応用／ベクトル解析と積分定理／微分方程式／複素数と複素関数／数値解析／確率，統計／金利計算／二進法と十六進法／公式集

F.R.スペルマン・N.E.ホワイティング著
東大 住　明正監修　前環境研 原澤英夫監訳

環境のための 数学・統計学ハンドブック

18051-0　C3040　　　　　Ａ５判 840頁 本体20000円

環境工学の技術者や環境調査の実務者に必要とされる広汎な数理的知識を一冊に集約。単位換算などごく基礎的な数理的操作から，各種数学公式，計算手法，モデル，アルゴリズムなどを，多数の具体的例題を用いながら解説する実践志向の書。各章は大気・土壌・水など分析領域ごとに体系的・教科書的な流れで構成。〔内容〕数値計算の基礎／統計基礎／環境経済／工学／土質力学／バイオマス／水力学／健康リスク／ガス排出／微粒子排出／流水・静水・地下水／廃水／雨水流

Ordinary Differential Equations and Physical Phenomena

原著日本語版

シリーズ〈新しい工学〉1

常微分方程式
と物理現象

神田　学 ［著］

常微分方程式の基礎的な解法から，
Java を用いた数値解析，非線形現象
までをやさしく解説．

B5 判 116 頁 本体 2300 円＋税

ISBN 978-4-254-20521-3

朝倉書店の英語書籍

Geography of Tokyo 　（『東京地理入門：東京をあるく，みる，楽しむ』英語版）

edited by

Toshio Kikuchi, Hiroshi Matsuyama, Lidia Sasaki, and Eranga Ranaweerage

A5 判 168 頁 本体 2800 円＋税

ISBN 978-4-254-16362-9

The Japanese Language（英語で学ぶ日本語学）1

Japanese Linguistics 　（日本語学）

by

Mark Irwin and Matthew Zisk

A5 判 304 頁 本体 4800 円＋税

ISBN 978-4-254-51681-4

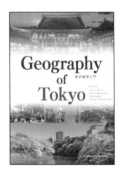

上記価格（税別）は 2020 年 10 月現在